T0213836

SpringerBriefs in History of Science and Technology

More information about this series at http://www.springer.com/series/10085

Matthias Schemmel

Historical Epistemology of Space

From Primate Cognition to Spacetime Physics

 Springer

Matthias Schemmel
Max Planck Institute for the History
   of Science
Berlin
Germany

ISSN 2211-4564   ISSN 2211-4572   (electronic)
SpringerBriefs in History of Science and Technology
ISBN 978-3-319-25239-1   ISBN 978-3-319-25241-4   (eBook)
DOI 10.1007/978-3-319-25241-4

Library of Congress Control Number: 2015951799

Springer Cham Heidelberg New York Dordrecht London

© The Author(s) 2016
This work is subject to copyright. All rights are reserved by the Publisher, whether the whole or part
of the material is concerned, specifically the rights of translation, reprinting, reuse of illustrations,
recitation, broadcasting, reproduction on microfilms or in any other physical way, and transmission
or information storage and retrieval, electronic adaptation, computer software, or by similar or dissimilar
methodology now known or hereafter developed.
The use of general descriptive names, registered names, trademarks, service marks, etc. in this
publication does not imply, even in the absence of a specific statement, that such names are exempt from
the relevant protective laws and regulations and therefore free for general use.
The publisher, the authors and the editors are safe to assume that the advice and information in this
book are believed to be true and accurate at the date of publication. Neither the publisher nor the
authors or the editors give a warranty, express or implied, with respect to the material contained herein or
for any errors or omissions that may have been made.

Printed on acid-free paper

Springer International Publishing AG Switzerland is part of Springer Science+Business Media
(www.springer.com)

Preface

From Primate Cognition to Spacetime Physics—What do the orientation of apes and the theory of relativity have to do with each other? This book discusses different forms of spatial thinking and their relation in a long-term history of knowledge. Starting from an analysis of the elementary structures of spatial knowledge found in animals and humans, it then investigates how human spatial knowledge is further shaped by various societal conditions. These conditions range from the universal human ability to share knowledge by means of language to the very specific development and ongoing differentiation of disciplinarily structured science. Other conditions relate to the emergence of systems of notation in the context of administrative challenges in the early civilizations, or to the systematic reflection on all sorts of knowledge in traditions of disputation. Scientific concepts of space such as Newton's absolute space or Einstein's curved spacetime are thus presented in their rootedness in pre-scientific knowledge structures. At the same time it is shown how these concepts are part of broader conceptual systems that are able to integrate expanding corpora of experiential knowledge.

This book should be viewed as a first attempt in the direction of a historical epistemology of space. Its main goal is to show that such a historical epistemology is possible at all: that the different forms of spatial knowledge are indeed related in their development, and that the study of this interrelated development may indeed provide insights into their epistemic status. The examples from different cultures and historical ages are chosen so as to substantiate systematic points. No attempt was made to give a balanced, let alone exhaustive, account of the world history of spatial thinking, since this is not the aim of the book. One can easily think of other examples from various cultures and times whose discussion under the perspective presented here would further contribute to the historical epistemology of space. Obvious examples are traditions of optics and perspective, and non-European traditions of theoretical reflection. It is therefore hoped that the book is taken as an inspiration to pursue further studies of this kind, which would not necessarily be restricted to spatial concepts. As will become clear from the arguments in this book, the epistemic separation of space from the context of other fundamental concepts such as time, matter, and force is, at least in part, itself due to historical circum-stances. Therefore, the immediate cognitive context of spatial thinking cannot be excluded if a full understanding of the long-term development of spatial knowledge is aimed for. Further contributions to a historical epistemology of space along similar lines will appear in: Matthias Schemmel (ed.), *Spatial Thinking and External Representation: Towards an Historical Epistemology of Space*. Edition Open Access, Berlin.

This work was prepared at the Max Planck Institute for the History of Science in Berlin within the context of the TOPOI project cluster, with the support of both institutions. Jürgen Renn and Peter Damerow† were of invaluable help in setting up, shaping, and realizing the project. For reading earlier versions of the text and for their helpful comments and discussions I would further like to thank Sascha Freyberg, Tal Glezer, Robert Goulding, Pietro Omodeo, John Stachel, Martin Thiering, and Matteo Valleriani. For careful reading and discussion of particular chapters, I am grateful to Wulf Schiefenhövel (Chap. 3), Jens Høyrup (Chap. 4), and Alexander Blum (Chap. 7). I would further like to thank Lindsay Parkhowell for his thoughtful language editing (the errors that remain are mine).

Keitum   Matthias Schemmel
July 2015

Contents

Chapter 1
The Challenge of a Historical Epistemology of Space

Abstract The chapter introduces the program of a historical epistemology of space and discusses the relation of the historical development of knowledge to its phylogenetic and ontogenetic developments. The chapter further provides an overview on the remaining chapters of the book.

Keywords Phylogenesis · Ontogenesis · Historical development · Knowledge · Cognition

In the history of Western epistemological thought there is a long tradition of dividing human knowledge into a purely rational part, independent of any experience in the outer world, and an experiential part. Traditionally, many aspects of spatial knowledge have been claimed to belong to the rational part. Prominent examples range from the Pythagorean-Platonic claims about the ideal existence of geometrical figures, via early modern rationalistic ideas of deriving properties of space from pure reasoning, to the axiomatic deduction of properties of space in the logical positivism of the early twentieth century and later constructivist philosophies.[1] In all these cases it is attempted, on very different grounds, to draw a clear-cut line between what is known of space prior to experience and the spatial knowledge that is derived from experience. Particularly influential was Immanuel Kant's description of space as a pure form of intuition. Theorems from geometry are among Kant's prime paradigms

---

[1] For Plato, see, for instance, the discussion on geometry in *Politeia*, 526c 9–527c 11. A prominent rationalistic treatment of space is found in René Descartes' *Principles of Philosophy*, Part 2, in particular §§ 8–21 (Descartes 1644, 37–44; for an English translation, see Descartes 1984, 42–49). An example of the division of spatial knowledge into an a priori and an experiential part from the early twentieth century is Carnap (1922, 62–67), who distinguishes formal, intuitive, and physical space, of which only the first is completely independent of experience; at the same time Carnap argues that the cognitive structure given by a topological space of infinitely many dimensions is the precondition for any kind of spatial experience. For a constructivist argument about the a priori nature of Euclidean space, see, for instance, Lorenzen (1984), who wants to show "how the Euclidean theorems are to be proven in Plato's sense solely from definitions and postulates (as fundamental constructions)." ("[…] wie die euklidischen Theoreme im Sinne Platons allein aus Definitionen und Postulaten (als Grundkonstruktionen) zu beweisen sind," Lorenzen 1984, 15, English translation M.S.)

for the existence of synthetic a priori judgements.[2] In his *Metaphysical Foundations of Natural Science*, Kant applies his program of isolating the a priori part of knowledge to the science of his time.[3]

The historical epistemology of space shares this interest of identifying the different sources of spatial knowledge. At the same time it is based on a thoroughly genetic, or developmental, view of cognition. According to this view, experiential knowledge participates in the construction of cognitive structures, which in turn constitute the basis for further experience. From this viewpoint a static separation between preformed structures of cognition and contingent experiences is impossible. Or rather, it is possible only in the snapshot image of a 'cognitive subject.' If the idea of a foundation of human knowledge—and scientific knowledge in particular—is justified, then this foundation must consequently lie in the reconstruction and analysis of the processes that have over the course of time led to this knowledge. Kant's program of exploring which aspects of our knowledge originate in preformed cognitive structures and which aspects involve empirical insights is thus transformed into one of studying the history of the interactive processes between experience and structures of knowledge. It is in this vein that the historical epistemology of space attempts to address the problem of the epistemic status of spatial knowledge by studying its history.

The developmental view of cognition is strongly suggested by results from different empirical disciplines. First and foremost, evolutionary biology teaches us that cognition is a function of the human organism, in particular the brain, and is therefore to be understood as a product of biological evolution. Furthermore, from studies in developmental psychology it has become clear that many fundamental cognitive structures are not present at the moment of a child's birth, but are only gradually built up over the years in the long process of growing up. Finally, studies in the history of science and philosophy have revealed the historicity of fundamental concepts such as *space*, *time*, *force*, and *matter*, a historicity that became most blatant through the radical changes associated with the rise of the theories of relativity and the quantum in early twentieth century physics.

Accordingly, one may distinguish three interwoven strands of development for which the role of experience in the process of building up the perception and conception of space can be studied: (1) the phylogenetic strand, i.e., the development of the biological species *Homo sapiens*; (2) the ontogenetic strand, i.e., the development of individual human beings; and (3) the historiogenetic strand, i.e., the development of human society and culture through history.

The phylogenesis of cognition is the subject matter of *evolutionary epistemology*. Continuity of development is produced by heredity. While experience pertains to individuals, given the background of genetic variation it shapes the species' development stochastically through its impact on an individual's ability to contribute its

---

[2]See Kant's *Transcendental Exposition of the Concept of Space* in his *Critique of Pure Reason*, B 41–42 (Kant 1998, 69–70).
[3]Kant (1997).

genes to the next generation's gene pool (i.e., through selection). In this way, since genes define a species' cognitive potential, a generation's experience has a bearing on the next generation's basis for experience and thus for further cognitive evolution.[4]

The ontogenesis of cognition is the subject of *genetic epistemology*. Continuity of development is produced by the identity of an individual's psyche. Experience may become part of the individual's memory and shape developing cognitive structures, which are mental reflections of real actions. The cognitive structures in turn constitute the basis for further action and related experience and, as a consequence, for further cognitive development.[5]

The historiogenesis of cognition is the subject of *historical epistemology*. Continuity of development is produced by external knowledge representations which serve the social reproduction of cognitive structures within a culture or their transfer between cultures. This reproduction relies on institutions structuring the use of the external representations.[6] Experiential knowledge is encoded in these external representations, which in turn become the precondition for further experience and the construction of new cognitive structures. These may then become encoded in higher-order representations which are the basis for further experience and further cognitive development.[7]

The historiogenetic strand is the one that will further concern us here.[8] It is closely interwoven with the other strands in two fundamental ways. First of all, in anthropogenesis, phylogenetic and historiogenetic factors are closely entangled. The emergence of human culture and with it the onset of the historical development of human cognition was a result of biological evolution and, as a consequence, necessarily built upon its biological foundations. But not only did human biology condition the onset of human culture, this culture also conditioned the last steps of anthropogenesis.[9]

The second way historiogenesis is related to the two other strands of cognitive development is based on the fact that the species' development, its phylogenesis as well as its historiogenesis, is realized through the individuals' ontogeneses. Thus, the phylogenesis of cognitive structures depends on the ontogenetic transformation

---

[4]See Lorenz (1977) for a classic work on evolutionary epistemology and Vollmer (1994) for a concise overview.

[5]See Piaget (1970) and other works by Piaget cited in this book.

[6]'Institutions' are here understood in the most general sense as social patterns structuring and controlling collective actions.

[7]Cf. Damerow (1996, 371–381). Accounts of historical epistemology as the term is understood here include, among others, Damerow (2007) and Renn (2004, 2005).

[8]Related studies are Damerow (2007) concerning the concept of number, and Dux (1992) and Elias (1988) concerning the concept of time. A programmatic outline of a historical epistemology of perception is Wartofsky (1979). For long-term histories of concepts of space in science and philosophy, see Gent (1971), Jammer (1954), and Gosztonyi (1976).

[9]See, for instance, Schurig (1976, in particular 164–214), for a discussion of the coevolution of anatomy and culture in anthropogenesis. For a more recent account and further references to the literature, see Odling-Smee et al. (2003, 239–281) who discuss coevolution from the perspective of niche construction.

of the genotypes into phenotypes, and only the latter are subject to natural selection. In a similar way, the historiogenesis of cognitive structures depends on individuals who appropriate the collective knowledge available in a given society at a given time in history in their ontogenesis and who, through their cognitive activities, participate in the transmission and transformation of this knowledge.

The entanglement of ontogenetic and historiogenetic developments of cognition explains the central role played by external knowledge representations for understanding long-term developments in the history of knowledge. The means of representation, such as communicative action, spoken language, artifacts, drawings, maps, writing and other symbol systems, mediate between socially shared knowledge, which is the subject of historical development, and the individuals' knowledge which, while being subject to all the contingencies of the individuals' biographies, is the only actual realization of the shared knowledge. While the external means of knowledge representation define a space of possible transformations of shared knowledge, such transformations actually occur only through the thinking of individuals, which is in turn conditioned by their participation in this knowledge.

The recognition of this dialectic between individual thinking and shared knowledge is crucial for an understanding of the aim of a historical epistemology of space as outlined here. This aim is not to give a narrative of the world history of individual acts of spatial thinking. Such an aim would not only be unachievable, owing to the sheer magnitude of the task, but also theoretically unsatisfactory, precisely because it neglects the social dimension of individual thinking. The aim is rather to describe historically identifiable and theoretically interpretable cognitive configurations, or stages, that demarcate the horizon of the forms of spatial thinking that are possible in a given historical situation.[10]

The identification of stages does not imply that the historical development of the forms of spatial thinking is a linear process, or even a process directed at an aim.[11] Rather, this development shares some qualitative features with biological evolution, even though it is governed by entirely different mechanisms. These features include:

- *Unpredictability of future developments*: Developmental processes are complex and interconnected, such that future developments are, as a rule, unpredictable at any time in history.
- *Dependency of later developments on earlier ones*: Despite this indeterminacy, earlier developments produce the necessary preconditions for later ones.

---

[10]Cf. Damerow (1994, 312).

[11]We may speak of development whenever change produces circumstances that serve as a necessary precondition for specific further changes. To deny historical development of cognition would mean to deny the dependency of cognition on its earlier forms and thus, ultimately, to deny its dependency on society and culture. But, as shall be argued in this book, this dependency is what distinguishes human cognition from animal intelligence. Its denial would mean to assume naively that any thought and insight was possible at any time in history. The idea of the historical development of cognition should not be taken as implying value-judgements, of course. The outright identification of developmental approaches with value-judgements reveals an (often unconscious) ethnocentrism, since it uncritically presupposes that 'our modern' modes of thinking are per se valued higher.

- *Temporal directedness of overall development*: This dependency of later developments on earlier ones explains aspects of the temporal order of development and makes it possible to define *earlier* and *later* stages of spatial thinking.
- *Asynchrony of development*: The temporal directedness does not imply, however, that all development proceeds uniformly on a global scale: different stages coexist and there may even be local or temporal developments from a 'later' stage to an 'earlier' one.

In the following chapters of this book, six different aspects of the historical development of spatial knowledge are discussed. Each chapter starts with a section delineating the object relevant for the study of the respective aspect. The following sections present one or more examples. Each chapter ends with a section characterizing the epistemic status of the spatial knowledge discussed in the chapter.

The similar biological constitution of all humans and the fundamental similarities in their physical environments make it plausible to assume that there are structures of spatial cognition that do not vary between different cultures or over history, but constitute the foundation for all cultural manifestations of spatial knowledge. In order to understand the dependence of spatial thinking on culture it is important first to identify these structures. Chapter 2 discusses the sensorimotor schemata that are formed in humans regardless of society and historical age in similar ways as with nonhuman primates. The examples presented are (1) the schema of permanent objects, which allows for successful handling of objects on a mesocosmic scale, and (2) the landmark model of larger-scale space underlying cognitive mapping skills and allowing for successful navigation through various types of environment.

The essential difference of human as compared to animal cognition is to be found in the social abilities of human beings. Humans possess unique abilities to share knowledge, a fact that constitutes the basis for the cultural evolution of human spatial cognition, leading to elaborate cultural systems for environmental orientation. Chapter 3 discusses mental models of large scale space that humans share by means of targeted actions, gestures, spoken language, and other kinds of material knowledge representation. Two well-studied examples from recent non-literate societies are presented: (1) the network of spatial designations of the Eipo of West New Guinea, and (2) the absolute-directional system of the expert navigators of the Caroline Islands of Micronesia. It is argued that the cultural practices build upon the elementary structures of spatial knowledge described in Chap. 2, which are, at the same time, modified and partly overridden by them.

A new form of spatial knowledge develops in early civilizations, in which the allocation and management of land necessitates an administrative control of space and leads to the formation of new means of knowledge representation. Chapter 4 discusses the transformation of human societies from bands and tribes to city-states and empires, which brought about new forms of the social control of space, involving techniques of surveying, writing, and drawing, which became the precondition for the development of geometry and thereby shaped the further development of spatial thinking. The example of Mesopotamia is presented, where practices of area determination are documented on clay tablets from the late fourth millennium BCE on.

In the following millennia the system of representations developed in the context of administrative and educational institutions. It is argued that this resulted in a metrization of cognitive models of space, albeit confined, at the time, to a small group of experts.

From ancient societies such as Greece and China there is evidence for theoretical reflections on the representations of elementary and practical forms of spatial knowledge. In both societies this development can be argued to be closely related to the emergence of cultures of disputation and a vivid tradition of writing. Chapter 5 discusses the knowledge emerging from such reflections as being distinguished from the elementary and practical forms on which it builds by its greater generality and aspiration for consistency. Two traditions are taken as examples, both originating in ancient Greece and both being further pursued, under different social and cultural circumstances, up into modern times: (1) the tradition of deductive geometry, which originated in the reflection on practical knowledge involving the use of drawing instruments; and (2) the tradition of philosophies of space, which originated in the reflection on the linguistic representation of elementary spatial knowledge.

Besides the reflection on representations of existing spatial knowledge, the expansion of spaces of experience is a motor for conceptual development, whether these are the geographical spaces known through political expansion, trade, and exploration, the cosmological spaces known through observation, or the microcosmic spaces known through engineering and experimentation. Chapter 6 presents three examples for processes of concept formation and the generalization of spatial concepts that were promoted by such expansions of experiential spaces. The first example refers to the systematic accumulation of geographical knowledge, which laid the foundation for the introduction of a global system of terrestrial coordinates. This allowed landmarks to be related no longer just to other landmarks but also to a mathematically determined, abstract geographical space. The second example refers to the accumulation over centuries of astronomical and mechanical knowledge, which, by a process of reflective integration, brought about the Newtonian concept of a homogeneous, isotropic, absolute space independent of its matter content. The third example relates to the expansion of knowledge of microscopic space by institutionalized research on electric and magnetic forces, which brought about and stabilized the concept of the electromagnetic field.

The renewed revolution of the concept of space in twentieth century physics can again be understood as a process of reflective knowledge integration, this time however an integration of disciplinarily highly structured knowledge. Chapter 7 discusses the loss of autonomy of the concept of space that resulted from the demise of the Newtonian concept. The examples presented are (1) the spacetime of special relativity, which emerged from an integration of knowledge from mechanics and electrodynamics and resulted in a close entanglement of the concepts of space and time; and (2) the spacetime of general relativity, which emerged from the additional integration of knowledge on gravitation and resulted in a close entanglement of the concepts of space and matter (or energy).

The concluding chapter summarizes some key insights of this study by high-lighting developmental strands that connect different forms of spatial knowledge. The chapter closes by positioning the presented approach within the larger field of knowledge studies, in particular arguing for its potential to consolidate two aspects which are widely perceived to work in opposite directions: the insight that knowledge depends on culture and history on one hand, and the aspiration for a rational foundation of knowledge on the other (Chap. 8).

References

Carnap, R. (1922). *Der Raum. Ein Beitrag zur Wissenschaftslehre*. Berlin: Reuther & Reichard.
Damerow, P. (1994). Vorüberlegungen zu einer historischen Epistemologie der Zahlbegriffsentwick-lung. In G. Dux & U. Wenzel (Eds.), *Der Prozeß der Geistesgeschichte. Studien zur ontogenetis-chen und historischen Entwicklung des Geistes* (pp. 248–322). Frankfurt a.M.: Suhrkamp.
Damerow, P. (1996). *Abstraction and representation. Essays on the cultural evolution of thinking*. Boston Studies in the Philosophy of Science (Vol. 175). Dordrecht: Kluwer.
Damerow, P. (2007). The material culture of calculation. A theoretical framework for a historical epistemology of the concept of number. In U. Gellert & E. Jablonka (Eds.), *Mathematisation and demathematisation. Social, philosophical and educational ramifications* (pp. 19–56). Rotterdam: Sense Publication.
Descartes, R. (1644). *Principia philosophiae*. Amstelodami (Amsterdam): Elzevirium.
Descartes, R. (1984). *Principles of philosophy*. Dordrecht: Reidel.
Dux, G. (1992). *Die Zeit in der Geschichte. Ihre Entwicklungslogik vom Mythos zur Weltzeit*. Frankfurt a.M.: Suhrkamp, mit kulturvergleichenden Untersuchungen in Brasilien (J. Mensing), Indien (G. Dux / K. Kälble / J. Meßmer) und Deutschland (B. Kiesel).
Elias, N. (1988). *Über die Zeit*. Frankfurt a.M.: Suhrkamp.
Gent, W. (1971). *Die Philosophie des Raumes und der Zeit. Historische, kritische und analytische Untersuchungen*. Hildesheim: Georg Olms.
Gosztonyi, A. (1976). *Der Raum. Geschichte seiner Probleme in Philosophie und Wissenschaften*. Freiburg: Alber.
Jammer, M. (1954). *Concepts of space: The history of theories of space in physics*. Cambridge, MA: Harvard University Press.
Kant, I. (1997). *Metaphysische Anfangsgründe der Naturwissenschaft*. Hamburg: Meiner.
Kant, I. (1998). *Kritik der reinen Vernunft* (1. & 2. Aufl.). Hamburg: Meiner.
Lorenz, K. (1977). *Behind the mirror. A search for a natural history of human knowledge*. London: Methuen.
Lorenzen, P. (1984). *Elementargeometrie. Das Fundament der Analytischen Geometrie*. Mannheim: Bibliographisches Institut.
Odling-Smee, F. J., Laland, K. N., & Feldman, M. W. (2003). *Niche construction: The neglected process in evolution*. Princeton, NJ: Princeton University Press.
Piaget, J. (1970). *Genetic epistemology*. New York: Columbia University Press.
Renn, J. (2004) The paradox of scientific progress. Notes on the foundation of a historical theory of knowledge. In Research Report 2002–2003 (pp. 21–49). Max Planck Institute for the History of Science.
Renn, J. (2005). The relativity revolution from the perspective of historical epistemology. *Isis, 95*(4), 640–648.
Schurig, V. (1976). *Die Entstehung des Bewußtseins*. Frankfurt a.M.: Campus.
Vollmer, G. (1994). *Evolutionäre Erkenntnistheorie* (6th ed.). Stuttgart: Hirzel.
Wartofsky, M. W. (1979). Perception, representation, and the forms of action: Towards an historical epistemology. *Models: Representation and the scientific understanding* (pp. 188–210). Dordrecht: Reidel.

Chapter 2
Natural Conditions of Spatial Cognition

Abstract The similar biological constitution of all humans and the fundamental similarities in their physical environments make it plausible to assume that there are structures of spatial cognition that do not vary between different cultures or over history, but constitute the foundation for all cultural manifestations of spatial knowledge. In order to understand the dependence of spatial thinking on culture it is important first to identify these structures. The chapter discusses the sensorimotor schemata that are formed in humans regardless of society and historical age in similar ways as with nonhuman primates. The examples presented are (1) the schema of permanent objects, which allows for successful handling of objects on a mesocosmic scale, and (2) the landmark model of larger-scale space underlying cognitive mapping skills and allowing for successful navigation through various types of environment.

Keywords Primate cognition · Spatial thinking · Cognitive mapping · Object permanence · Sensorimotor intelligence

The Object of Study

In order to understand how human spatial thinking depends on the cultural conditions present at different times in history it is of fundamental importance first to identify spatial abilities and corresponding cognitive structures that are *not* products of human culture, and accordingly not subject to historical change. These may be termed the *natural conditions of spatial cognition*. Starting from such an identification one may then ask how historical and present-day cultural manifestations of spatial thinking relate to this universal basis.

The natural conditions of spatial cognition are of a double origin. First, there are biological predispositions of the human species which also involve a cognitive dimension. Second, there are features of the physical environment in which each individual grows up that are so fundamental that they are independent of culture. In the first case, it is the mechanisms of biological evolution by which experience enters the formation of cognitive structures, in the second it is each individual's experience in ontogenesis. The two origins are closely entangled, however, since the ontogenetic

© The Author(s) 2016
M. Schemmel, *Historical Epistemology of Space*, SpringerBriefs
in History of Science and Technology, DOI 10.1007/978-3-319-25241-4_2

unfolding of biological predispositions always takes place in a physical environment which exhibits certain universal features. While the question of the relation between the two origins shall not further concern us here, it is important to note that the idea of universal aspects in human spatial cognition does not in itself imply any kind of nativism.[1]

When trying to identify the natural conditions of spatial cognition we encounter a methodological problem. Cross-cultural studies help to identify aspects of spatial thinking that are human universals, i.e., aspects that do not depend on the particularities of any specific culture (for instance on the use of a particular language); yet, the universal aspects identified in this manner will include aspects that depend on the very existence of human culture (for instance on the presence of language altogether). From their birth on (and in certain respects even before that), human beings are immersed in their culture. They are born into a cultural *habitus* that shapes their social and physical experiences and thus potentially exerts an influence on their cognitive development. More importantly, they participate in specifically human modes of cultural learning.[2] As a consequence, when studying the ontogenesis of human cognition, it is practically impossible to abstract from processes of the individual's enculturation. Therefore, to unveil its natural conditions, human spatial cognition has to be compared to animal cognition as the cognition of beings without human culture. Of particular interest in this context is the cognition of nonhuman primates, since cognitively they appear closest to humans and are probably similar to our not-yet-human ancestors. The natural conditions of human spatial cognition arguably comprise their spatial abilities and the corresponding cognitive structures.[3]

To identify the natural conditions of spatial cognition the object of study must therefore be the spatial behavior of animals and humans (children and adults), and in particular of nonhuman primates. Of central relevance in this context are the abilities of *object permanence* and *cognitive mapping*. In the following, these abilities shall be described and explained in terms of their implications concerning the fundamental structures of human spatial cognition.

Example: Object Permanence

Object permanence is what developmental psychologists call the mental construction of objects as entities independent of the self, which are understood to exist in a definite location or move along a definite trajectory in space. Studies in developmental psychology suggest that what may be called the *schema* of permanent object is not

---

[1] For a critical discussion of 'nativist' approaches, see, e.g., Tomasello (1999, 48–51).

[2] For an explanation of cultural *habitus*, see Tomasello (1999, 78–81); for that of cultural learning, see Tomasello (1999, 61–70), who relates these human modes of learning to the conception of others as intentional beings and argues that its development begins around the ninth month.

[3] For a more critical discussion of comparisons between animal and human spatial cognition, see Hazen (1983).

present at the time of a child's birth, but only develops during the first two years of childhood.[4]

The construction of the schemata of object on the one hand and space on the other are indissociable, as space can only be constructed concurrently with objects and vice versa.[5] No change in our perceptions could be understood as a change of place or position of something if there were no unchanging objects. On the other hand, for something to be an object it must necessarily occupy a certain space, i.e., be at a certain location and have a certain shape and size. To be able to understand certain changes as arising from one's own motion relative to the objects, it is furthermore necessary to conceive of one's own body as being positioned in a common space with the objects.

Following Piaget one can distinguish six stages of sensorimotor development, which describe the progressive dissociation of the objects and their spatio-temporal trajectories from the subject's activities.[6] We will particularly focus on the last three stages.

At the beginning of the development 'space' is heterogeneous in the sense that the spatial aspects of different senses and actions (oral space, visual space, auditory space, tactile space, the space of body positions, etc.) are not coordinated and thus not integrated into one structure. When the ability to grasp what is seen develops, this leads to the construction of schemata of action under visual control and the perceptual constancy of shape and size. The changes in the perception of bodies in motion (or changes perceived when the subject's body is in motion) are no longer understood as transformations of the 'objects', but rather as changes in perspective. A feeding-bottle, to mention one of Piaget's examples, is turned around by the infant in order to find the rubber teat, indicating the construction of the permanent object, with all of its parts being conserved.[7] Changes in body position are then gradually differentiated from changes of state. These developments can be considered the beginning of object permanence.

At this stage, however, infants seek a hidden object where they last found it and not where they saw it vanish.[8] Piaget interprets this finding to the effect that the object is still only a part of a situation characterized by the successful action, i.e., there is no object independent of action or at least no continuous trajectory of a body

---

[4]For a definition of the concept of schema, see, for instance, Piaget (1983, 180–185). A different definition is given in Neisser (1976, 51–57). Below we will introduce the concept of *mental model* to describe relevant cognitive structures.

[5]See Piaget (1959, in particular 97–101).

[6]Various empirical studies have been devoted to testing Piaget's theory of cognitive development, complementing and correcting it in many respects. But while some of the interpretations have been at great variance with Piaget's views, the evidence does not seem to refute Piaget's overall scheme as outlined here. For a review of much of the literature and a critical discussion of post-Piagetian work on spatial cognition, see Newcombe and Huttenlocher (2003).

[7]For example Piaget (1981, 110–111).

[8]This behavior is often referred to in the psychological literature as the *A-not-B error*, 'A' denoting the location where the object was previously found and 'B' the one at which it vanished; see, e.g., Piaget (1981, 109–110) and Newcombe and Huttenlocher (2003, 53–71).

in space.[9] This stage of the construction of objects, usually referred to as *stage four*, and reached by human infants towards the end of their first year, may therefore for our purposes be called the *stage of action-bound locations*.

At the next stage (*stage five*), infants start proactive enquiry and systematic observation. They seek a hidden object based on the perceived displacement, i.e., they no longer search for it where they last found it, but rather where they saw it vanish. They are not able, however, to infer the position of an object that has been moved outside their view. This stage, reached by human infants at about the beginning of their second year, is here called the *stage of perception-bound locations*.

Stage six, by contrast, may be referred to as the *stage of perception-independent locations*. It is reached when the infant systematically seeks for a hidden object and does so exclusively at locations to which the object can possibly have moved. For instance, when an object is moved under a cover along a row of boxes and put into one of them, stage six ability means that the infant seeks for the object only in those boxes into which the investigator can possibly have put it. This ability therefore involves the mental representation of the displacement or trajectory of an object even if it cannot be seen while it is being moved.

In adult humans the abilities that indicate full development of object permanence appear to be universal. While there are indications that the speed of development varies, not only between individuals but between whole cultures, there are no studies known to me that would deny this ability for the members of any culture. Most studies do in fact take these abilities for granted.[10]

Object permanence skills have further been proven for many animal species.[11] In various studies it has been tested whether nonhuman primates and other mammals are able to locate objects when they have witnessed how they were hidden and if they can infer the location of an object that has changed place outside their view. A survey of the studies indicates that all primates and at least some nonprimate mammals (cats and dogs in particular) possess stage four and five skills of object permanence, i.e., they apprehend perception-bound trajectories.[12] To test whether primates possess stage six skills and apprehend perception-independent trajectories, an object is hidden, e.g., in a small box, which is then moved under several covers. From the observed seeking behavior of the animal it can then be concluded whether it can infer the possible location of an object that has been moved outside its view. Stage six skills have been proven rigorously for only a few primate individuals. Among nonprimate mammals, dogs possess these skills, while cats fail to do so.[13]

---

[9]For a detailed discussion of this stage, see Piaget (1959, 44–66).

[10]See the discussion in Dasen and Heron (1981, 303–307).

[11]For a survey of the spatial abilities of nonhuman primates, see Tomasello and Call (1997).

[12]Tomasello and Call (1997, 41–42).

[13]According to Tomasello and Call (1997, 46), many studies which suggest stage six skills have not employed appropriate control procedures. One may speculate that the occurrence of stage six abilities depends on the specific needs of an animal species, e.g., when following prey or when avoiding predators (Tomasello and Call 1997, 55).

There is thus clear evidence that the sensorimotor schemata of object permanence in general, and the ones underlying stage six abilities in particular, are not unique to humans. On this basis one may argue that they belong to the natural conditions of human spatial cognition.

Action and perception under the control of the schemata of object permanence imply spatial structures among which there are the following:

- *Dichotomy of objects and spaces*: Objects are tangible (albeit not always reachable), and between them there are non-tangible (i.e., 'empty') spaces.
- *Definiteness and exclusivity of place*: Every object is in a place and always in one place at a time. No other object can be in the same place at the same time.
- *Three-dimensionality of objects and spaces*: Objects are extended in such a way that different sides of an object are perceptible from different perspectives. There is a concealed backside of each object (like the feeding bottle's rubber teat in Piaget's example). The spaces between objects are likewise extended, allowing for objects not only to be located side by side, but also to obstruct the view to another object.
- *Distinction of vertical direction*: There is one direction distinguished by the tendency of most objects (including one's own body) to fall down or to resist lifting.
- *Continuity of object trajectories*: The mutual spatial relations of objects, including one's own body, may change, i.e., there is motion. The trajectories of motions are continuous, i.e., there are no 'jumps': objects do not vanish in one place and re-appear in another, but pass through all intermediate places during the motion. Stage six abilities indicate that the schema of permanent object implies continuous trajectories regardless of whether they are perceived or not.

Example: Cognitive Mapping

Besides the smaller-scale skills related to object permanence, humans develop sophisticated abilities of spatial orientation on larger scales. They can quickly accumulate spatial information about previously unknown territories; in known territories they can move flexibly, i.e., they can make detours and take short cuts that they have not previously made or taken; and they can optimize their routes by arranging the stations of their travel in a rational manner. They can integrate knowledge about landmarks with knowledge about the motion of their own body to construct route knowledge, and combine their knowledge about intersecting routes to obtain what may be called configurational knowledge: knowledge about the overall configuration of landmarks and their relations.[14] They are further able to make use of cues such as wind directions, the position of the sun, or distal landmarks. Following a large body of literature, these abilities are here referred to as *cognitive mapping*.[15]

---

[14]Siegel and White (1975); Kitchin and Blades (2002, 89–90).

[15]See Kitchin and Blades (2002) for a recent account on cognitive maps which surveys a large part of this literature.

Evidence for cognitive mapping in human adults is abundant. Members of hunter-gatherer societies as well as inhabitants of modern cities construct cognitive maps and use them in their everyday orientation and navigation, even though the concrete techniques and abilities of spatial orientation vary widely over different societies and different ecologies.[16]

As is the case for object permanence, cognitive mapping skills are subject to ontogenetic development.[17] They appear to develop later than the former, though, which seems plausible given the limited mobility of young infants and also the fact that the use of landmarks arguably presupposes object permanence of at least stage four (action-bound locations). The later development of cognitive mapping makes it even more difficult than was the case with object permanence to disentangle the natural development of biological predispositions from processes of the individual's enculturation and assess the scope of natural conditions of spatial cognition within cognitive mapping. Again, the comparison with animal cognition may provide important clues.

Besides humans, various species of animals exhibit sophisticated performance in spatial orientation.[18] Striking spatial abilities are found in various species of birds and mammals, in particular also rodents. It was, in fact, in the context of studies on the orientation abilities of rats in mazes that the term *cognitive map* was coined. Edward Tolman and his collaborators convincingly demonstrated that the rats' behavior could not be fully accounted for solely by means of stimulus and response. Rather the rats' spatial memory of the maze had to be organized in such a way that they could draw inferences about alternative pathways they had not employed before.[19]

Dogs and other mammals have also been shown to be able to use detours and shortcuts.[20] Nonhuman primates, in particular, have been shown to be able to use spatial information in a flexible manner.[21] Chimpanzees, for instance, who were shown how food was hidden at several locations in a familiar environment were later able to retrieve most of the food, whereby they did not follow the order in which the food was placed, but an order that reflected a minimum-effort strategy. They could also be shown to first retrieve, using such a strategy, the kinds of food they preferred before they proceeded to less preferred food.[22] Hamadryas baboons, to give another example, were observed remembering the locations of important sites such as sources of water in their local environment, to use least distance strategies in their travel, and even to speed up when approaching a known site well before they could have perceived it, thus demonstrating that they knew where they were.[23]

---

[16]See, e.g., Hazen (1983).

[17]See, e.g., Kitchin and Blades (2002, 85–96).

[18]See various contributions in Pick and Acredolo (1983).

[19]Tolman (1948).

[20]See, for instance, Fabrigoule (1987).

[21]See Tomasello and Call (1997, 28–39) for a survey of the evidence for different primate species.

[22]See Menzel (1973). Menzel (1987) discusses the interpretation of these findings in terms of cognitive mapping.

[23]Sigg and Stolba (1981).

We can summarize these findings to the effect that the basic human cognitive mapping skills—just as object permanence skills—are not indicative of a peculiarity of human cognition but are among its natural conditions[24]:

> Overall, primates have the general mammalian spatial skills of cognitive mapping and object permanence [...]. [...] It is also unlikely that humans have any special skills in these domains of spatial cognition. They too possess the general mammalian skills of cognitive mapping and object permanence (with clear stage 6 skills early in ontogeny) .

The cognitive mapping skills imply fundamental spatial structures such as the following:

- *Dichotomy of movable and unmovable objects*: Some objects can be moved or move by themselves (e.g., conspecifics); other objects cannot be moved, i.e., they have a fixed location (e.g., trees). These latter objects thus define a ground against which one's own motion and the motion of other objects is perceived.
- *Focus on plane of movement*: The space of movement (structured by a network of landmarks, places, and regions) mostly lies within a more or less horizontal plane. (The additional importance of the vertical depends on the mode of life in particular ecologies such as living on different levels of a forest, of a mountainous region, or of a city with multi-story buildings.)
- *Path-connectedness of plane of movement*: The topology of the plane of movement is path-connected, i.e., between any two locations there is a path connecting them (otherwise it would not be a plane of movement). Generally, there may be different paths along which one may arrive at the same location and one may travel along a closed path and come back to one's initial location, even in cases where the path encircles obstacles that cannot be overcome (e.g., trees, mountains, river sections, or buildings).
- *Dependency of effort on path taken*: The effort it takes to get from one location to another generally depends on the path taken.

The Character of Spatial Knowledge

What is the epistemic status of the natural conditions of spatial cognition? It has been argued here that these conditions are rooted in sensorimotor intelligence, which is characterized by a close relation between cognition and concrete action.[25] The development of sensorimotor activity, roughly spanning the first two years of human

---

[24]Tomasello and Call (1997, 55–56). There are further studies pointing to similarities in animal and human spatial cognition. Thus, Foreman et al. (1984), who carried out experiments with preschool children in a so-called radial maze, an arrangement previously used in experiments on spatial abilities of animals, have pointed to remarkable similarities between pre-school children and well-trained nonhumans in the performance of certain spatial tasks. This fact was interpreted to suggest a similarity of the role of visuospatial cues in the development and use of cognitive representations of space and the underlying processes across species.

[25]See Piaget (1981, 107–116); Piaget (1959, 86–96); Piaget and Inhelder (1956, 5–13).

life, involves the reflexes, grows to include habits, and culminates in the emergence of practical intelligence. In the course of this development, sensory data are assimilated to cognitive structures called *schemata of action*, which are in turn accommodated to the increasing amount of sensorimotor experience. The result is an increasing coordination, generalization, and differentiation of schemata of action which constitute human sensorimotor intelligence.[26]

It is important to note that the implied spatial structures described in the two preceding sections are not themselves objects of thought. They allow for successful action, but there is no indication that the related spatial abilities imply any consciousness, i.e., any reflection upon the schemata controlling the actions, and thereby go beyond the sensorimotor realm.[27] Thus, without the *dichotomy of objects and spaces*, no object could be perceived or grasped. Without the *dichotomy of movable and unmovable objects* no stable mental representation of the environment was possible. Without the *three-dimensionality of objects and spaces* no change in the visual image could be understood as a change of perspective. But while these structures allow for spatial inferences to be drawn, they do so only in the context of action and perception and are otherwise inaccessible to the actor.[28] This becomes clear, for example, when school children who successfully find their way from home to school and back are unable to represent these routes in a map-like fashion.[29] Another example is provided by the well-attested difficulties of children to rotate a landscape in their minds and describe how it would look from a different point of view.[30]

In particular, there is no indication of symbol use or the dependence of spatial cognition on external knowledge representations in general.[31] Accordingly there are also no concepts of space. The cognitive structures forming the natural conditions of spatial cognition common to all humans do not represent general, or abstract, ideas,

---

[26]See, e.g., Piaget (1981). See also Damerow (1998, 248).

[27]They rely on what Piaget has called *perceptional space* in distinction to *representational space*, which is built up only at the preoperational and operational stages (Piaget and Inhelder 1956, 3–43). See, however, Boesch and Boesch (1984, 168–169) who interpret some of their findings as evidence for concrete operational thinking in the spatial reasoning of nonhuman primates and suggest the existence of 'Euclidean' cognitive maps, relating to Piaget's distinction between topological, projective, and Euclidean space; see also Normand and Boesch (2009).

[28]It remains an open question to what extent the transfer of spatial abilities to novel and artificial contexts of action presupposes that the actor's understanding of the novel situation is one involving a representation of real space. For example, it may be doubted whether the fact that rhesus macaques, using a joystick, are able to anticipate the path through a computer-simulated maze (see Tomasello and Call 1997, 51–54) necessarily implies that they conceive of the maze as a representation.

[29]Piaget (1960, 3–26).

[30]See the classical experiment by Piaget and Inhelder (1956, 209–246). For a critical discussion integrating recent empirical results, see Newcombe and Huttenlocher (2003, 118–125).

[31]A possible counter example of symbol use in spatial communication among bonobos is discussed in Savage-Rumbaugh (1998, 161–165), but does not seem conclusive.

but depend on the specific contexts of action and perception. They are not to be found on the level of concepts but on that of the schemata controlling sensorimotor behavior.[32]

Besides the notion of schema of action we shall employ the concept of *mental model* in referring to these cognitive structures. The concept of mental model refers to internal knowledge representation structures that allow current experience to be processed by relating it to former experience. The structure of the model consists of *slots* and their mutual relations. The slots are filled by specific instances, i.e., by an input from the current situation fulfilling certain conditions required by the slot. But these slots may also have default fillings which are effective whenever appropriate current information is not available. The default fillings of slots is one way earlier experience is encoded in the mental model. In fact, the very structure of the model is a result of earlier accommodations to experience.[33] In this way, a mental model allows the perception of, understanding of, or even reasoning about, a situation whenever the situation can be assimilated to the model successfully—even in cases where the available information is incomplete. A major reason to introduce the concept of mental model here, and not simply speak of sensorimotor schemata, is that mental models function on different levels of cognition. The sensorimotor and practical mental models inform the models functioning on higher conceptual and theoretical levels (and these may in turn have repercussions on the lower levels).[34]

The sensorimotor *mental model of a permanent object* is a mental structure into which sensory data are assimilated when objects are perceived and handled. For the assimilation to be successful, the shape, size, location, and position of the object must be identifiable. But they do not need to be constant in time. The sensorimotor schemata that underlie the model ensure that certain changes in perception are interpreted as changes of perspective, i.e., of the position of the object or one's own body in respect to it, rather than as changes of the object itself. As becomes clear from our discussion above, the sensorimotor model in its fully developed form further implies the mental representation of continuous trajectories.

To describe a range of abilities in large-scale spatial orientation, we have employed the term *cognitive mapping*. This term is widely used, but the precise character of the mental representation underlying the related abilities is a matter of controversy. In particular, it is not at all clear that this representation can be characterized as a bird's eye view of the environment as the term 'map' suggests. Just as the mental model of

---

[32] We reserve the notion of concept to describe elements of knowledge structures that are somehow related to linguistic or otherwise symbolic representations, without implying, of course, that there was a one-to-one relation between concepts and words.

[33] For an introduction to the concept of *accommodation* (the adaption of mental structures to environmental inputs) and the complementary concept of *assimilation* (the adaption of environmental inputs to mental structures; see below), see Piaget (1981, 7–9 and *passim*).

[34] On the concept of mental model as understood here, see in particular Renn and Damerow (2007); see also various contributions in Gentner and Stevens (1983). The concept is akin to Marvin Minsky's *frames* (Minsky 1975).

object does not presuppose a three-dimensional mental image,[35] the mental representation of the large-scale environment need not take the form of a two-dimensional map.[36]

Here the corresponding cognitive structures shall again be described in terms of mental models. The *mental models of large-scale space* may be conceived of as networks of landmarks and their spatial interrelations. It is plausible to assume that the landmarks and their relations are part of a hierarchical structure, in which places and regions of different size are defined by reference to landmarks or other places and regions.[37] The landmarks, places, and regions are further endowed with contextual information about what is found there, e.g., kinds of food, water, predators and conspecifics, tools, and places to rest. The spatial relations between landmarks, places, and regions of different size involve topological information (inclusion, order along a route, proximity) as well as information on distances and angles. This latter information is given not in terms of numerical measures, of course, but rather in terms of sensorimotor experiences about variation in travel effort, about viewing directions to landmarks, and about perspectives. Configurations of landmarks, places, and regions can further be related to reference points outside the realm of motion such as the sun or distal landmarks like a large mountain, or to directions defined by features within the local environment such as a slope of the landscape or recurring winds. The landmarks that fill the model's slots are permanent objects or configurations of such objects, so that the elementary knowledge about objects in general (their permanence, their change of appearance with perspective and distance, etc.) applies to them. The structural relations between the slots contain the knowledge about the spatial relations among the landmarks. The individual realizations of the mental models of large-scale space are highly dependent on the concrete features of the respective environment, since they encode the experiential knowledge accumulated as the individual moves through this environment. Nevertheless, the basic structure applies universally. This universal structure will in the following be referred to as the *landmark model of space*.

---

[35]It is the functioning of the model—for instance the way different perspectives are coordinated to make an object remain constant in size and shape under different views—that implies the three dimensionality. For a suggestion of how a three-dimensional cube and its transformations under different perspectives may be realized mentally without invoking a three-dimensional mental image, see Minsky (1975, 216–221), who uses coordinated *frames*. A more comprehensive discussion of three-dimensional vision is found in Marr (1982).

[36]Objections against imputations of the use of cognitive maps, in particular when simpler explanations of the spatial abilities are available, are raised, for instance, by Tuan (1975) and Bennett (1996). Recently, Wang and Spelke (2002) argued against the concept of cognitive map, emphasizing the human use of navigation techniques such as path integration, which are also found in insects and spiders and imply no more than the mental representation of one vector. It seems, however, that the presence of more 'momentary' and 'egocentric' representations does not at all preclude the build-up of more enduring and comprehensive mental representations. On the relation of these two types of representations, see, for instance, Cornell and Heth (2004).

[37]See Gärling et al (1985) for a detailed description of the possible components cognitive maps are made of.

References

Bennett, A. T. D. (1996). Do animals have cognitive maps? *The Journal of Experimental Biology*, *199*, 214–224.

Boesch, C., & Boesch, H. (1984). Mental map in wild chimpanzees: An analysis of hammer transports for nut cracking. *Primates*, *25*(2), 160–170.

Cornell, E. H., & Heth, C. D. (2004). Human spatial memory: Remembering where. In G. L. Allen (Ed.), *Memories of travel: Dead reckoning within the cognitive map* (pp. 191–215). Mahwah, NJ: Erlbaum.

Damerow, P. (1998). Piaget, evolution, and development. In J. Langer & M. Killen (Eds.), *Prehistory and cognitive development* (pp. 247–269). Mahwah, NJ: Lawrence Erlbaum Associates.

Dasen, P. R., & Heron, A. (1981). Cross-cultural tests of Piaget's theory. In H. C. Triandis & A. Heron (Eds.), *Handbook of cross-cultural psychology, Vol. 4: Developmental psychology* (pp. 295–341). Boston: Allyn & Bacon.

Fabrigoule, C. (1987). Study of cognitive processes used by dogs in spatial tasks. In P. Ellen & C. Thinus-Blanc (Eds.), *Cognitive processes and spatial orientation in animal and man, Vol. 1: Experimental animal psychology and ethology* (pp. 114–123). Dordrecht: Martinus Nijhoff Publishers.

Foreman, N., Arber, M., & Savage, J. (1984). Spatial memory in preschool infants. *Developmental Psychobiology*, *17*(2), 129–137.

Gärling, T., Böök, A., & Lindberg, E. (1985). The development of spatial cognition. In R. Cohen (Ed.), *Adults' memory representations of the spatial properties of their everyday physical environment* (pp. 141–184). Hillsdale, NJ: Erlbaum.

Gentner, D. & Stevens, A. (Eds.). (1983). *Mental models*. Hillsdale, MI: Lawrence Erlbaum Associates.

Hazen, N. L. (1983). Spatial orientation: A comparative approach. In H. L. Pick Jr & L. P. Acredolo (Eds.), *Spatial orientation: Theory, research, and application*. New York: Plenum Press.

Kitchin, R., & Blades, M. (2002). *The cognition of geographical space*. New York: Tauris.

Marr, D. (1982). *Vision: A computational investigation in the human representation of visual information*. San Francisco, CA: Freeman.

Menzel, E.W. (1987). Behavior as a locationist views it. In Ellen, P. & Thinus-Blanc, C. (Eds.), *Cognitive processes and spatial orientation in animal and man, Vol. 1: Experimental animal psychology and ethology* (pp. 55–72). Dordrecht: Martinus Nijhoff Publisher.

Menzel, E. W. (1973). Chimpanzee spatial memory organization. *Science*, *182*(4115), 943–945.

Minsky, M. (1975). The psychology of computer vision. In P. H. Winston (Ed.), *A framework for representing knowledge* (pp. 211–277). New York: McGraw-Hill.

Neisser, U. (1976). *Cognition and reality: Principles and implications of cognitive psychology*. San Francisco, CA: Freeman.

Newcombe, N. S., & Huttenlocher, J. (2003). *Making space: The development of spatial representation and reasoning*. Cambridge, MA: MIT Press.

Normand, E., & Boesch, C. (2009). Sophisticated Euclidean maps in forest chimpanzees. *Animal Behaviour*, *30*, 1–7.

Piaget, J. (1959). *The construction of reality in the child*. New York: Basic Books.

Piaget, J. (1981). *The psychology of intelligence*. Totowa, NJ: Littelfield Adams & Co.

Piaget, J. (1983). *Biologie und Erkenntnis*. Frankfurt: Fischer.

Piaget, J., & Inhelder, B. (1956). *The child's conception of space*. London: Routledge & Kegan Paul.

Piaget, J., Inhelder, B., & Szeminska, A. (1960). *The child's conception of geometry*. Abingdon: Routledge.

Pick, J., Herbert, L. & Acredolo, L.P. (Eds.). (1983). *Spatial orientation: Theory, research, and application*. New York: Plenum Press.

Renn, J., & Damerow, P. (2007). Mentale Modelle als kognitive Instrumente der Transformation von technischem Wissen. In H. Böhme, C. Rapp, & W. Rösler (Eds.), *Übersetzung und Transformation* (pp. 311–331). Berlin: de Gruyter.

Savage-Rumbaugh, S. (1998). Scientific schizophrenia with regard to the language act. In J. Langer & M. Killen (Eds.), *Piaget, evolution, and development* (pp. 145–169). Mahwah, NJ: Erlbaum.

Siegel, A. W., & White, S. H. (1975). The development of spatial representation of large-scale environments. In H. W. Reese (Ed.), *Advances in child development and behavior*. New York: Academic Press.

Sigg, H., & Stolba, A. (1981). Home range and daily march in a hamadryas baboon troop. *Folio Primatologica, 36*, 40–75.

Tolman, E. C. (1948). Cognitive maps in rats and men. *The Psychological Review 55*(4), 189–208.

Tomasello, M. (1999). *The cultural origins of human cognition*. Cambridge, MA: Harvard University Press.

Tomasello, M., & Call, J. (1997). *Primate cognition*. New York: Oxford University Press.

Tuan, Y. F. (1975). Images and mental maps. *Annals of the Association of American Geographers, 65*(2), 205–214.

Wang, R. F., & Spelke, E. S. (2002). Human spatial representation: Insights from animals. *Trends in Cognitive Sciences, 6*, 376–382.

Chapter 3
Culturally Shared Mental Models of Space

Abstract The essential difference of human as compared to animal cognition is to be found in the social abilities of human beings. Humans possess unique abilities to share knowledge, a fact that constitutes the basis for the cultural evolution of human spatial cognition, leading to elaborate cultural systems for environmental orientation. The chapter discusses mental models of large scale space that humans share by means of communicative actions, gestures, spoken language, and other kinds of material knowledge representation. Two well-studied examples from recent non-literate societies are presented: (1) the network of spatial designations of the Eipo of West New Guinea, and (2) the absolute-directional system of the expert navigators of the Caroline Islands of Micronesia. It is argued that the cultural practices build upon the elementary structures of spatial knowledge described in the previous chapter, which are, at the same time, modified and partly overridden by them.

Keywords Social cognition · Spatial orientation · Eipo · Caroline Islands · Navigation

The Object of Study

If the natural conditions of human spatial cognition are similar to those of some animal species, as has been argued in the previous chapter, what then accounts for the obvious distinction of human spatial abilities and thinking? Rather than attributing this distinction to some specifically human biological disposition for *spatial* cognition, the point shall be made here that the distinction can be explained as resulting from uniquely human abilities of *social* cognition. One argument against the existence of a specifically human *module* for spatial cognition is based on considerations of the necessary time scales for processes in biological evolution.[1] Another argument would be obtained once it were shown that the specificity of human social cognition, together with the historical development of human thought ensuing from it, can satisfactorily explain the characteristics of human spatial cognition such that no further

---

[1] See Tomasello (1999, 54–55).

© The Author(s) 2016
M. Schemmel, *Historical Epistemology of Space*, SpringerBriefs
in History of Science and Technology, DOI 10.1007/978-3-319-25241-4_3

biological factors have to be invoked. Substantiating this claim is a key task for a historical epistemology of space.

The human ability of social cognition implies that humans are able to communicate, to share knowledge, and to learn from each other. For this kind of cognition to come about it is crucial that humans understand their conspecifics as intentional beings, i.e., as beings who act purposefully just like themselves, and to be able to imagine themselves in another's place.[2] In order to communicate about space, human children must learn to take the perspective of others. For this they have to construct a mental representation of space that allows them to conceive all possible perspectives. This means the construction of what Piaget calls *representational space* as distinguished from *perceptional space*.[3] It is the social aspect of human cognition that implies representations that go beyond those closely tied to action and perception occurring at the latest stages of sensorimotor development.[4]

Sharing knowledge crucially depends on what Piaget calls the *symbolic function*, i.e., the ability to distinguish events and objects from their meaning. In human ontogeny this ability emerges at the preoperative stage, which succeeds the sensorimotor stage. On its basis, actions of conspecifics can be understood to mean something, i.e., they become potential means of knowledge representation. Purposeful actions with the aim of communicating knowledge, like gestures, and directed joint action become possible. Tools likewise come to represent knowledge in relation to the actions performed with them. Another particularly powerful means of knowledge representation and communication is human language, which phylogenetically is assumed to have developed in the course of the Paleolithic.[5] Visual representations like drawings are also known to have existed in Paleolithic times. They are attested to by various kinds of extant artifacts, most prominently the cave paintings of the Upper Paleolithic. In the course of continued cultural evolution, the very means of external knowledge representation develop further, as may be exemplified by the emergence of writing and the use of other sign and symbol systems such as numerical notation under the particular socio-cultural circumstances of early city-states (see Chap. 4).

The crucial distinction between animal and human cognition, then, is the emergence of a cumulatively evolving human culture, a thoroughly social phenomenon. For all the abilities of individual humans that may arguably play a crucial role in the emergence of this culture, such as the abilities to use and produce tools, to understand conspecifics as intentional beings, and to understand symbols and develop language, we find precursors in the animal kingdom.[6] Rather than being attributable to a single distinguishing factor, the animal–human divide seems to emerge from a process

---

[2]On specifically human ways of learning, see Tomasello et al. (1993) and Tomasello (1999, 26–55).

[3]Piaget and Inhelder (1956, 3–43).

[4]Piaget (1959, 364–376).

[5]Referring to results from neurology, developmental psychology, and archaeology, it has been speculated that the development of human language was closely related to the communication of cognitive maps (Wallace 1989).

[6]Besides Tomasello and Call (1997), see, for instance, the discussion of cognitive abilities such as categorization as developing independent of language in Langer (2001) and reports on tool-

in which social, material, and cognitive developments interact in a complex causal structure.[7]

An immediate consequence of the cultural evolution of human societies on spatial cognition is that the mental models of large-scale space become culturally shared. In addition to those commonalities between two individuals' mental models of space that are due to their similar biological constitutions and their similar experiences within the same environment, human mental models of space display cultural commonalities. In this way the mental models of space themselves become part of an evolving culture, accumulating collective experience over generations and thus exceeding in richness and refinement any mental model a single individual could have produced.

The sharing of mental models of space appears to be common to all human societies, from nomadic tribes to modern urban societies. When considering the impact of the cultural sharing of knowledge on the mental models of large-scale space, the general objects of study are therefore the practices of navigation and spatial orientation in all kinds of human societies. In most contemporary societies, however, these practices involve specialized means of spatial representation and advanced technology which have developed over the long course of history. To get an idea of what is achievable in the absence of maps, compasses, sextants, or GPS receivers, one has to study the spatial practices of nonliterate societies that do not employ such specialized material tools. In the case of prehistoric societies, the archaeological evidence is the only available source for a reconstruction of such practices. In the case of recent nonliterate societies, by contrast, spatial practices can be investigated much more directly, which makes them a preferred object of study.

Recent nonliterate societies show a wide variety of cultural systems for spatial orientation and communication.[8] This cultural diversity is due not only to the self-referential dynamics of cultural evolution, but obviously also to the fact that these systems represent responses to the challenges of very different ecologies to which they are adapted. Nevertheless, there are common patterns that may be discerned. It may be observed, for instance, that toponyms play a central role for spatial reference in a wide range of societies. Places and their relations are richly endowed with meanings relating to mythology, the history of places, and the natural knowledge about them. In many societies this practice is furthermore complemented with a system of absolute directions which in some cases play such a crucial role that members learn always to keep track of these directions.[9]

Two examples highlighting these different aspects shall be sketched here, the network of toponyms and spatial reference of the *Eipo* living in the central highlands of West New Guinea, and the absolute-directional system of Micronesian navigators.

(Footnote 6 continued)
making and tool-using abilities and linguistic capacities of Bonobo individuals (Schick et al. 1999; Savage-Rumbaugh and Fields 2000).

[7]See, for instance, Damerow (2000) and Jeffares (2010).

[8]See, e.g., Burenhult (2008), Senft (1997), Levinson and Wilkins (2006).

[9]Various examples are given in Levinson and Wilkins (2006).

Example: The Eipo's Network of Toponyms and Spatial Reference

When referring to space, the Eipo, rather than using measures of distance and direction, employ a close-knit network of toponyms.[10] The geographical area in which they live and move around is densely covered with the names of mountains, caves, valleys, rivers, confluences of rivers, villages, gardens, meadows, lakes, rocks, peculiar trees, and so forth. But landmarks are not only used in order to refer to certain places: spatial reference in itself plays a central role in Eipo culture. Space is a fundamental principle for ordering and classifying objects, and is used by the Eipo for their taxonomy of plants and animals. Eipo space is furthermore structured by mythical assignments with pervasive consequences for social life. The division of land by borders pervades the whole space known to the Eipo, which ends where there are no relatives or trade partners. This space traditionally had a radius of about three day's marches before the Eipo became increasingly acculturated in recent decades. Among the first deeds of the ancestors was the division of land and its distribution among the clans. Salient landmarks are considered to be traces of the totem ancestors who granted the land to the tribe. Spirits who inhabit the area around such landmarks may do harm to people if the landmark is destroyed.

The men's house is the place of assembly and habitation for all men and initiated boys of an Eipo village. All other relevant structures of the village are arranged more or less concentrically around this central point of reference. But a men's house of an Eipo community is not only the center of the habitat, it is also a sacred place. According to the Eipo founding myths, human society began when the forefathers built a men's house. When one of the ancestors said: 'I will build a men's house here, you build over there', this signified the occupancy of a certain territory. Sacred objects preserved in the men's houses serve as a legitimatory expression of continued territorial claims. Ritual actions in the context of the men's house symbolize the claim on the surrounding area by maintaining connection to the time of arrival at the site. Of particular significance for the demarcation of territory is the *yurye* (*Cordyline terminalis*), a small tree which is considered sacred. Besides its use at sacred places, e.g., around the men's house, it is also used to mark off garden land in someone's possession. The borders of a garden are given by imagined lines connecting the *yurye*. Transgressing the border of another individual or family's garden is understood as a violation of the sacred order.

---

[10]On the Eipo, see Schiefenhövel (1991). The Eipo's spatial practice and language is discussed in more detail in Thiering and Schiefenhövel (forthcoming). The Eipo are arguably the best-documented among recent nonliterate societies with a material culture comparable to that of certain Neolithic societies. The systematic studies on them were published in the series *Mensch, Kultur und Umwelt im zentralen Bergland von West-Neuguinea*, Berlin: Reimer, and in a series of films published by IWF. The following exposition is based on Eibl-Eibesfeldt et al. (1989, 30–33 and 140–143).

The non-metrical character of Eipo space also finds reflection in language.[11] There seems to be no general term for *distance* in the Eipo language. There are terms for keeping a distance when walking (*tekisib-* and *karen*), and the distance to another village may be described by the number of nights that pass on the way there, so that a journey of more than two days' marches becomes "a journey on which I slept twice" (*betinye mamse bisik*). Practices of measuring lengths or distances have been observed only very rarely and when measurements were performed by anthropological researchers they seemed quite unnecessary to the Eipo.[12] Further, there seems to be no term for a general concept of angle, which could be used in designating directions, although there is quite a general term for something spread or forked (*kwa*) and there are terms for the angle between stem and shoot (*keila*) or between upper and lower leg (*keil buseling*).

Example: The Absolute-Directional System of the Caroline Island Navigators

The well-documented example of navigation between the Caroline Islands presents the case of a highly elaborate mental model of space which allows for safe navigation between remote islands without the help of any instrument specially designed for the purpose of spatial orientation.[13] The mental model makes use of absolute directions and dead reckoning and is therefore quite different from that of the Eipo. The corresponding knowledge was handed down from generation to generation mainly on the basis of joint action and oral instruction with very little reinforcement from visual representations (mostly transitory drawings produced in the sand). Its representations thus remain within the realm of a Neolithic material culture.[14]

The navigators of the Caroline Islands of Micronesia, who constitute a specially educated expert group within their society, travel on their boats between islands so remote that land is sometimes out of their sight for several days. Among the difficult

---

[11] The following language examples are taken from Heeschen and Schiefenhövel (1983).

[12] Michel (1983, 13). There is one instance of an Eipo practice reported (and filmed) that may document a kind of measurement in terms of a direct comparison of lengths. After having arranged the posts for the construction of a men's house in a circular manner, a roll of bark, which was later to become the floor cover, was apparently used to estimate the appropriateness of the arrangement (Koch and Schiefenhövel 1979; Koch 1984, 53). Another instance, the measurement of the width of a landing strip by means of a rope, is described in Thiering and Schiefenhövel (forthcoming). Notice, however, that the possibility of the anthropologists' measuring activities influencing the behavior of the Eipo cannot be excluded in this context.

[13] An early report on the navigation techniques on the Caroline Islands was given by Safert (1911). On the navigators of the Puluwat Atoll and their techniques of spatial orientation, see in particular Gladwin (1974). For a reconstruction of the cognitive functioning of these techniques, see Hutchins (1983). To what extent the described practices have survived to the present day is not known to me.

[14] Gladwin (1974, 125–143)

tasks they master is "to tack upwind to an unseen target keeping mental track of its changing bearing [...]."[15] The navigators' knowledge comprises specificities of highly local character which tell them where they are, e.g. swell patterns in the vicinity of land, wind and weather patterns, and the presence and behavior of sea birds. It furthermore contains a detailed knowledge of certain stellar constellations referred to as *the Micronesian star compass*, and the relations between islands with respect to these constellations.

The star compass, the centerpiece of Micronesian navigation, is an elaborate expert version of the practice, common to most cultures, to identify and communicate directions by reference to shared environmental or celestial cues. Often these cues are given by the motion of heavenly bodies, such as the sun (sunrise, sunset), a star near the pole, the slope of a mountain (uphill, downhill, and transverse), or seaward and landward directions in regions near a coastline. In distinction to these practices, the Micronesian star compass defines 32 directions.

The compass consists of 14 stellar paths, with each path composed of several stars that seem to follow one after the other in their nocturnal path across the sky. Depending on the time of night, one of these stars indicates an eastward and another a westward direction. Since the Caroline Islands are only about 8° north of the equator, the stars rise and set nearly perpendicular to the horizon, so that a star can easily be used to indicate a point on the horizon even if it has risen well above it. Together with the pole star and three different positions of the Southern Cross, this accounts for the 32 directions clearly discernible by night for those who acquired the star compass. According to Hutchins, "[...] a practiced navigator can construct the whole compass mentally from a glimpse of only one or two stars near the horizon."[16] At daytime the directions can be maintained by observing major ocean swells or the directions at which sun and moon rise and set and comparing them to the internalized star compass.

One main function of the star compass is to determine the direction from one island to another. Therefore, for each pair of islands the navigators learn the constellation under which they have to sail in order to get from one to the other. But the star compass is also used for tracking distances. To know how much of their path between two islands they have covered, the navigators track the position of a third island, the so-called reference island, under the star compass (sometimes they even use more than one reference island). Given their knowledge about the constellations under which the reference island is seen at the beginning and the end of their voyage, its position gives them a direct measure of the distance travelled. This measure is conceived in terms of time: taking into account the time of departure and knowing the usual travel time of the entire voyage, the reference island's position under the star compass is directly related to the time of the day (or night).

---

[15]Hutchins (1983, 192).
[16]Hutchins (1983, 195).

This use of reference islands as landmarks works even when the islands are out of sight, as is often the case, and even when the islands are wholly imaginary. The navigators track the reference island's position under the star compass in their minds. Somewhat speculatively, but remaining entirely within the mental model reconstructed for the Micronesian navigators, Hutchins is even able to explain how the navigators adjust the relation of travel time to the position of the reference island confronting varying travel speeds, and also how they tack against the wind without losing track of the position of the destination island and the remaining travel time.[17]

These navigational techniques amount to a mental transfer from dead reckoning to the use of movable landmarks (the islands) whose changing position is described relative to an absolute frame (the star compass). The navigators' shared mental model of the space through which they travel is a cultural construct clearly building upon the foundations of the sensorimotor cognitive structures of space encountered in the preceding chapter: it combines the use of dead reckoning with a network of landmarks and implies the *path-connected topology* of space, *focusing on the horizontal plane of movement*. It does not, however, imply a map-like bird's eye view of the area in which the moving canoe is placed. In particular, there is no evidence for the mental image of bundles of lines of sight running straight from successive positions of the boat on its line of travel to different positions on the star compass, all intersecting on the reference island. On the contrary, such a view seems to interfere with the mental model of the navigators. Hutchins refers to reports that reveal how difficult it was for the navigators to switch from their mental model to such a map-like view. Rather, in their mental model, the islands move and change position with reference to the star compass. To form the idea that the reference island lies on the intersection of two lines of sight would mean to envision oneself to be at two places at once![18]

The *dichotomy of movable and unmovable objects* is modified by this mental model of space in an interesting way. The star compass, i.e., the stars and stellar constellations *modulo* their nocturnal motions, is considered fixed. Since no motion of the canoe with respect to this fixed ground can be perceived, the canoe, too, is conceived of as being fixed in location. The islands thus become moving landmarks, but their motion is coordinated in such a way that the direction of the reference island is a direct indicator of the portion of travel time passed. This example illustrates that, while the sensorimotor models constitute a basis for the cultural development of mental models, specific features of the elementary model can be overridden by the development of a higher level of thought.

---

[17] Hutchins (1983, 220–223).

[18] Thus one of the navigators "eventually succeeded in achieving the mental tour de force of visualizing himself sailing simultaneously from Oroluk to Ponape and from Ponape to Oroluk […]. In this way he managed to comprehend the diagram [showing the place of the reference island at the intersection of the lines indicating its direction from Oroluk and Ponape, respectively]." (Lewis 1972, 143, as quoted in Hutchins 1983, 207).

The Character of Spatial Knowledge

The spatial knowledge described above may be characterized as practical knowledge. Among its characteristic features are: its *transmission through external knowledge representations*; its *cultural organization*; its *dependence on the specific contexts of action*; and its *locality*.

Transmission through external knowledge representations. In contrast to sensorimotor knowledge, which is built up in the individual's interaction with the physical world, practical knowledge is built up through social interaction and communication. The knowledge representations employed in this context include actions with the explicit aim of teaching, spoken language, and material means, such as—in the case of Micronesian navigation—arrangements of pebbles in the sand. While the communication builds upon shared sensorimotor structures, the use of external knowledge representations makes it possible to accumulate knowledge that could never be acquired through one individual's experience.

Cultural organization. This accumulation is accompanied—and, in fact, the mastery of the accumulated knowledge only made possible—by the cultural organization of knowledge. This organization implies both the institutional and the cognitive dimensions. The social reproduction of knowledge relies on more or less stable social patterns (institutions) structuring the collective use of the means of knowledge representation. This social organization and its material means are further correlated to the cognitive organization of knowledge. Thus, culturally shared large-scale space is spanned not only by landmarks, places, regions, and their relations, but by the meanings attached to these entities. These meanings organize the spatial knowledge and are given in the form of nomenclatures, narratives, or sets of practices.[19] In contrast to the sensorimotor mental models of space, large parts of this mental representation may be accessed deliberately by its holder, particularly in order to communicate about space.

Dependence on the specific contexts of action. The spatial concepts structuring practical knowledge are as a rule not abstract or general but depend on the specific contexts of action. They are not applications of more general concepts in concrete situations but are rather conditioned by these situations. Thus, the Micronesian star compass, which is used only in the context of navigation, does not depend on a general concept of angle. The straight lines along which the space between two islands is traversed—as long as tacking can be avoided—are not the result of a reflection on the shortest line between two points, as in Euclidean geometry. They

---

[19]Place and spatial order play an important role in Eipo myths and, conversely, mythical narratives are instrumental in handing down spatial knowledge (Heeschen 1990). This appears to be a widespread means for organizing spatial knowledge; another example are the practices of the Ngatatjara who live in the Australian desert and use myths and ritualistic sequences of events to memorize and communicate the cultural knowledge about their habitat. A brief description is given in Heth and Cornell (1985, 232–235). For a recent account of the use of songlines as "oral maps" by the Wardaman and other Aboriginal cultures, including further references to the literature, see Ray P. Norris and Bill Yidumduma Harney, *Songlines and Navigation in Wardaman and other Aboriginal Cultures* (http://www.atnf.csiro.au/people/Ray.Norris/papers/n315.pdf, accessed 7 April 2015).

rather result from the practice of orientation which prescribes for each pair of start and target islands the constellation under which to travel. This is just an extension of the sensorimotor schema to approach a visible object by moving towards it directly with the object substituted by a culturally communicated direction marker. Further, how the concepts structuring practical thinking about large-scale space relate to more small-scale spaces remains largely undefined. As a consequence, metrization remains fragmentary. Distances counted in days of travel are not brought into any relation with cubits or other length measures which may be employed on a different scale. In accordance with this, the Micronesian navigators convert the positions of reference islands directly into portions of travel time passed without taking a detour via some general unit measure of distance.

Locality. The shared mental models of large-scale space are local in character. Practical concepts of space depend on the particulars making up space, and are thus not generally applicable to arbitrary environments. Systems of toponyms, for instance, obviously apply only locally, since they inherit the dependence on the particular environment from the landmarks and relations they refer to. The same holds for most variable cues such as the swell-patterns used by Micronesian navigators and the pairs of islands and their directions under the star compass. But more structural elements of the system may be dependent on local peculiarities as well. Thus, the star compass works for the Caroline Island navigators only owing to their proximity to the equator, since only there do the stars and constellations rise and set nearly perpendicularly to the horizon. In regions more remote from the equator, the horizontal shifting of the stars would render the application of the star compass very difficult if not impossible.

Summing up, culturally shared mental models of large-scale space may be understood as collective elaborations and modifications of sensorimotor models. Just like these sensorimotor models they are based on the landmark model of space, from which they inherit many structural features. At the same time, they encode a larger body of experiential knowledge than the sensorimotor models: they integrate the experiential knowledge about the environment not only from one individual but from whole societies over the course of many generations. This integration is achieved by means of the cultural organization of knowledge, which necessarily reflects features of the local environment and displays cultural characteristics. Elementary knowledge structures thus serve as a foundation for culturally shared practices without determining their cognitive dimension. At the same time, culturally transmitted knowledge may have repercussions on the more elementary level of sensorimotor knowledge. Action and perception based on a culturally shared mental model may become intuitive, for instance when, as we have seen, knowledge of the star compass becomes internalized.

References

Burenhult, N. (Ed.). (2008). *Language and Landscape: Geographical Ontology in Cross-Linguistic Perspective.* Special Issue of Language Sciences (Vol. 30). Amsterdam: Elsevier.

Damerow, P. (2000). How can discontinuities in evolution be conceptualized? *Culture and Psychology, 6*(2), 155–160.

Eibl-Eibesfeldt, I., Schiefenhövel, W., & Heeschen, V. (1989). *Kommunikation bei den Eipo. Eine humanethologische Bestandsaufnahme im zentralen Bergland von Irian Jaya, West-Neuguinea, Indonesien.* Mensch, Kultur und Umwelt im zentralen Bergland von West-Neuguinea (No. 19). Berlin: Reimer.

Gladwin, T. (1974). *East is a big bird. Navigation & logic on Puluwat Atoll* (2nd ed.). Cambridge, MA: Harvard University Press.

Heth, C. D., & Cornell, E. H. (1985). A comparative description of representation and processing during search. In H. M. Wellman (Ed.), *Children's Searching: The Development of Search Skill and Spatial Representation* (pp. 215–249). Hillsdale, MI: Lawrence Erlbaum Associates.

Heeschen, V. (1990). *Ninye bún. Mythen, Erzählungen, Lieder und Märchen der Eipo (im zentralen Bergland von Irian Jaya, West-Neuguinea, Indonesien).* Mensch, Kultur und Umwelt im Zentralen Bergland von West-Neuguinea (Vol. No. 20). Berlin: Reimer.

Heeschen, V., & Schiefenhövel, W. (1983). *Wörterbuch der Eipo-Sprache. Eipo-Deutsch-Englisch.* Mensch, Kultur und Umwelt im zentralen Bergland von West-Neuguinea (No. 6). Berlin: Reimer.

Hutchins, E. (1983). Understanding Micronesian navigation. In Erlbaum, D. Gentner & A. L. Stevens (Eds.), *Mental models* (pp. 191–225). Hillsdale, NJ: Erlbaum.

Jeffares, B. (2010). The co-evolution of tools and minds: Cognition and material culture in the hominin lineage. *Phenomenology and the Cognitive Sciences, 9,* 503–520.

Koch, G. (1984). *Maligdam. Ethnographische Notizen über einen Siedlungsbereich im oberen Eipomek-Tal, zentrales Bergland von Irian Jaya, West-Neuguinea, Indonesien.* Mensch, Kultur und Umwelt im Zentralen Bergland von West-Neuguinea (No. 15). Berlin: Reimer.

Koch, G. Schiefenhövel, W. (2009). Eipo (West-Neuguinea, Zentrales Hochland) - Neubau des sakralen Männerhauses in Munggona. DVD, Produktionsjahr: 1974, first published in 1979; IWF Bestellnummer/Bandzählung E 2475.

Langer, J. (2001). The mosaic evolution of cognitive and linguistic ontogeny. In M. Bowerman & S. C. Levinson (Eds.), *Language acquisition and conceptual development* (pp. 19–44). Cambridge, MA: Cambridge University Press.

Levinson, S. C., & Wilkins, D. (Eds.). (2006). *Grammars of Space.* Cambridge, MA: Cambridge University Press.

Lewis, D. (1972). *We the navigators.* Honolulu, HI: The University Press of Hawaii.

Michel, T. (1983). *Interdependenz von Wirtschaft und Umwelt in der Eipo-Kultur von Moknerkon. Bedingungen für Produktion und Reproduktion bei einer Dorfschaft im zentralen Bergland von Irian Jaya, West-Neuguinea, Indonesien.* Mensch, Kultur und Umwelt im zentralen Bergland von West-Neuguinea (No. 11). Berlin: Reimer.

Piaget, J., & Inhelder, B. (1956). *The child's conception of space.* London: Routledge & Kegan Paul.

Piaget, J. (1959). *The construction of reality in the child.* New York: Basic Books.

Safert, E. (1911). Zur Kenntnis der Schiffahrtskunde der Karoliner. *Korrespondenz-Blatt der Deutschen Gesellschaft für Anthropologie, Ethnologie und Urgeschichte, 42,* 131–136.

Savage-Rumbaugh, S., & Fields, W. (2000). Linguistic, cultural and cognitive capacities of bonobos (pan paniscus). *Culture and Psychology, 6,* 131–153.

Schiefenhövel, W. (1991). Eipo. In T. E. Hays (Ed.), *Encyclopedia of world cultures, Vol. 2: Oceania* (pp. 55–59). Boston: G. K. Hall & Co.

Schick, K. D., Toth, N., Garufi, G., Savage-Rumbaugh, E. S., Rumbaugh, D. M., & Sevcik, R. A. (1999). Continuing investigations into the stone tool-making and tool-using capabilities of bonobo (pan paniscus). *Journal of Archaeological Science, 26,* 821–832.

Senft, G. (Ed.). (1997). *Referring to space: Studies in Austronesian and Papuan languages*. New York: Clarendon Press.

Tomasello, M. (1999). *The cultural origins of human cognition*. Cambridge, MA: Harvard University Press.

Tomasello, M., & Call, J. (1997). *Primate cognition*. New York: Oxford University Press.

Tomasello, M., Kruger, A. C., & Ratner, H. H. (1993). Cultural learning. *Behavioral and Brain Sciences, 16*(03), 495–511.

Thiering, M., & Schiefenhövel, W. (forthcoming). Spatial concepts in non-literate societies: Language and practice in Eipo and Dene Chipewyan. In M. Schemmel (Ed.), *Spatial thinking and external representation: Towards an historical epistemology of space*. Berlin: Edition Open Access.

Wallace, R. (1989). Cognitive mapping and the origin of language and mind. *Current Anthropology, 30*(4), 518–526.

Chapter 4
Social Control of Space and Metrization

Abstract A new form of spatial knowledge develops in early civilizations, in which the allocation and management of land necessitates an administrative control of space and leads to the formation of new means of knowledge representation. The chapter discusses the transformation of human societies from bands and tribes to city-states and empires, which brought about new forms of the social control of space, involving techniques of surveying, writing, and drawing, which became the precondition for the development of geometry and thereby shaped the further development of spatial thinking. The example of Mesopotamia is presented, where practices of area determination are documented on clay tablets from the late fourth millennium BCE on. In the following millennia the system of representations developed in the context of administrative and educational institutions. It is argued that this resulted in a metrization of cognitive models of space, albeit confined, at the time, to a small group of experts.

Keywords Surveying · Field measurement · Early civilizations · Ancient Mesopotamia · Babylonian mathematics

The Object of Study

One immediate consequence of the cultural evolution of human societies on spatial cognition, which was discussed in the previous chapter, is the development of elaborate practices of spatial orientation based on shared mental models of large-scale space. Another way in which the cultural evolution of human societies shapes spatial thinking rests upon the fact that the organization of society implies the social control of space. How is space divided among different individuals and social groups, what is the social function of different places, what are the places for public, sacred, or private affairs, who is allowed to go where, and who is allowed to use what land or even owns it? Questions of this kind can be observed arising in the context of the organization of any human society

The means for the social control of space depend on the respective form of social organization. In the case of small rural communities such as that of the Eipo described

© The Author(s) 2016
M. Schemmel, *Historical Epistemology of Space*, SpringerBriefs
in History of Science and Technology, DOI 10.1007/978-3-319-25241-4_4

in the preceding chapter, we may speak of a *mythical control of space*. Under the mythical control of space, knowledge about the social function of different places and about the allocation of space is largely represented by myths, which also ensure its social implementation. Despite the central role the division of land plays in social life, the mythical control of space does not provide standardized tools for measuring lengths and distances or for determining the quantitive measure of an area. The Eipo's construction of a sacred men's house, which has to be of a defined size and shape, for example, is a complex task which is mastered without recourse to material representations of spatial knowledge such as measuring rods, drawings, or any kind of specialized geometric language. The spatial knowledge necessary to build the house is instead embodied in the ritual actions specific to the Eipo culture.[1] The distribution of the garden lands among the Eipo is governed by clan-membership, heredity, and the capacity to cultivate the land. There are practices for delimiting fields (the demarcation of land by sacred Cordyline trees), but not for determining or estimating field sizes. Conflicts over the right to use a piece of land may lead to hostilities or be solved by negotiation, but their resolution never involves measurement.[2]

Historically, the earliest evidence for the systematic use of standardized measures for the social control of space stems from the so-called early civilizations. The growing population of Neolithic sedentary communities in some areas of the world was accompanied by the development of increasingly specialized food production, irrigation, and food storage technologies, and resulted in the emergence of stratified societies that controlled progressively larger spaces. The formation of city-states and larger empires brought about new phenomena in human culture such as central-ized administration, property regimes, monumental architecture, centralized religion, and new forms of standardized means of knowledge representation. In particular, it brought about new forms of the social control of space which may be referred to as the *administrative control of space*. These forms involved techniques of measuring, surveying, writing, and drawing, which implied a progressive metrization of space and led to a kind of proto-geometry.

A decisive strand in this bundle of developments was the emergence of new forms of the division of labor. Besides gender-specific forms of division of labor (consider hunting as a predominantly male activity, for instance) or practice-specific forms (as is the case for the experts of Micronesian navigation), a fundamental division became socially manifest: the division of physical and intellectual forms of labor. A physical and an intellectual component may be generally discerned in the human practices of using and producing tools. Concrete action is preceded by planning, i.e., selecting tools, determining the sequence in which they are used, and

---

[1] Koch and Schiefenhövel (2009), Koch (1984, 49–54). See also Thiering and Schiefenhövel (forth-coming).

[2] Wulf Schiefenhövel, personal communication. See also Michel (1983). Other instances of the mythical control of space may be identified in the spatial practices and the spatial thinking reported for the Bororo of the Brazilian central plateau—see the account of the socio-spatial structure of the village Kejara given by Lévi-Strauss (1955, 244–277)—and for the Temne in northern Sierra Leone (Little John 1963).

coordinating work in cases where more than one individual is involved. The growing complexity of the planning and organizational tasks in the stratified societies of the early civilizations now led to a division of labor along this intellectual-physical divide. The result was a specialization of intellectual labor, which became manifest in the emergence of professions such as the scribe, the administrator, and the surveyor,[3] and an administrative hierarchy reflecting the emergence of mental activities that coordinated other mental activities.[4]

As these mental activities are themselves dependent on material tools, the development of early civilizations went along with fundamental innovations of the means of external knowledge representation. This holds in particular for activities related to the social control of space such as architecture, urban planning, surveying, and field measurement which involved means of semantic and numerical notation as well as tools for graphical representation such as the compass and the ruler. Among the early civilizations in which such techniques developed are those of Mesopotamia, Egypt, China, and Meso- and South-America. The developments are well-documented in the case of Mesopotamia, where proto-writing emerged at the latest around 3200 BCE on the durable medium of cuneiform tablets, so that a large amount of administrative records are preserved.[5] Evidence for similar developments in ancient Egypt and China is more indirect. In the Egyptian case we have depictions of surveyors at work, e.g., the wall painting in the tomb of Menna in Thebes,[6] and mathematical texts on the calculation of areas such as parts of the Rhind Papyrus, but no administrative documents on the determination of field areas are preserved.[7] In the Chinese case evidence comes from much later times and again does not document early administrative practices.[8] In the case of the pre-Columbian civilizations of Meso- and South-America we see similar parallel developments of societal challenges like architecture and urban planning on one hand and of the means of external knowledge representation, such as the Incan Quipu or Mayan writing, on the other. However, because we mostly lack records documenting the concrete symbolical procedures

---

[3]These professions are documented in administrative sources of ancient Mesopotamian cities; they are, e.g., explicitly mentioned in texts from the city of Šuruppag (modern Fara) dating from around 2540 BCE, see Robson (2008, 31). In the proto-literate period, many of these functions were fulfilled by the temple-managers, see Høyrup (1994, 55).

[4]Damerow and Lefèvre (1996, 396–397). See also Renn and Valleriani (2014, in particular 13–18).

[5]We here follow the middle chronology, for which the third dynasty of Ur roughly coincides with the 21st century BCE and Hammurabi's reign dates from 1792 to 1750 BCE.

[6]See, e.g., Lyons (1927).

[7]The earliest evidence for the use of measuring ropes in Egypt is probably the Egypt numeral '100', which has the shape of a coil of rope and is attested in the Second Dynasty (ca. 2890 – ca. 2686 BC) (see Clagett 1992–1999, 752), but probably dates from an earlier time. I am grateful to Jens Høyrup for pointing this out to me.

[8]Consider, in particular, the *Jiu zhang suan shu* (*Nine Chapters on Arithmetical Techniques*), containing, among other things, problems on the calculation of field areas and probably dating to the first century CE. (Guo 1993, 79–213. For editions in European languages, see Vogel 1968; Shen et al. 1999; Chemla and Guo 2004.)

employed in their administrative practices, a reconstruction of these practices as comprehensive as in the Mesopotamian case appears impossible.[9]

The object of study concerning the emergence and early development of the administrative control of space, and the related gradual metrization of space, is therefore the practices of measuring and surveying, in particular in the early civilizations in Mesopotamia, Egypt, and China, but also in other ancient societies. In the case of Mesopotamia, the development from early practices of surveying to Babylonian geometry spans millennia and involves fundamental innovations such as the invention of the sexagesimal place value number system. In the following section, a few aspects of this development, and the structures of spatial knowledge implied by them, shall be described.[10]

Example: Field Measurement and the Metrization of Space in Ancient Mesopotamia

The administrative practices of the Mesopotamian surveyors of the third millennium BCE involved tasks such as the demarcation and measuring of fields, and the calculation of areas, and probably of other magnitudes such as quantities of seed and yields.[11] To accomplish this they used standardized measures for lengths and areas. These measures were partly derived from the traditional measuring tools and brought into systematic relations. The resulting system of units was further expanded according to the needs of the administration.

The material means the administrative activities were based on comprised standardized measuring tools for determining lengths as well as symbolic means for noting measures, calculating areas, and producing a record of the corresponding administrative act. For instance, taking the term *éš* into consideration, which was used as a unit of length for about 60 m but originally designates a rope, it appears that measuring lines were a common tool for determining length.[12] The material basis of the symbolic notation were signs indented onto clay tablets. The signs themselves can be described as proto-arithmetic signs and proto-writing, which, over the course

---

[9]On Aztec, Inca, and Maya city planning see, for instance, the respective entries in Selin (1997). On the difficulty to reconstruct the mathematical knowledge involved, see Ascher (1986), Vinette (1986), and other contributions in Closs (1986).

[10]Surveying and the determination of field sizes in ancient Mesopotamia and the emergence of Babylonian geometry are discussed in more detail by Damerow (forthcoming); see further Damerow (2001), Høyrup (2002), and Robson (2008).

[11]Damerow (2001, 247).

[12]See the entry for *éš* [*rope*] in the electronic Pennsylvania Sumerian Dictionary at http://psd. museum.upenn.edu/epsd/nepsd-frame.html. Accessed February 7, 2012.

of the third millennium, turned into elements of a full-fledged (glottographic) writing system on one hand and arithmetic notation on the other.[13]

Areas were calculated by decomposing fields into quadrangles and triangles and adding the corresponding areas. For triangles, half the product of their shorter sides were used as a measure of their area. When opposite sides of a quadrangle were of different length, half their sum was used as a basis to calculate the area. In the case of an irregular quadrangle of sides a, b, c, and d, this amounts to an application of the so-called *surveyors' formula*, according to which, in modern notation, the area is given by $(a + c)/2 \cdot (b + d)/2$.[14] These procedures do not presuppose a concept of angle as an object of mensuration[15] and, as long as the shapes of the figures involved are sufficiently regular (nearly rectangular quadrangles and nearly right-angled triangles), the method produces reasonably accurate results.

The oldest clay tablets documenting the calculation of measured areas date back to around 3,000 BCE. In the course of the third millennium, the commission and the methods of the surveyors remained basically unaltered, but there were quantitative changes concerning the size and complexity of fields and the accuracy of results. One consequence was the increased use of drawings of the fields. These field plans were not drawn to scale, but, being augmented with the inscription of numbers next to the lines, perfectly fulfilled the practical purpose of conveying the geometry of the problem.

The administrative practices also brought about intellectual activities more indirectly related to the actual tasks of the administration. There is a genre of texts which do not serve any obvious administrative purpose and which are generally characterized as *schooltexts*. They probably stem from the context of handing down the scribes' knowledge from generation to generation, an assumption that could explain their partial deviation from the immediately practical procedures.[16]

It was probably in the context of such intellectual activities that the sexagesimal place value system developed; earliest evidence for its existence stems from the time of the third dynasty of Ur (late third millennium).[17] The sexagesimal positional system made it possible to relate all metrological systems to a unified numerical system, in particular also the geometrical ones, and to apply the same general pro-

---

[13]See Damerow (2012, 170). For a discussion of the concept of glottographic writing, see Hyman (2006).

[14]This way of determining areas remained in use through millennia; it is documented much later in Egypt (Neugebauer 1934, 123), and was also used by the Roman *agrimensores* (Folkerts 1992, 324). The method may have been used independently by Aztec surveyors; for a reconstruction and discussion of Aztec methods of determining area, see Williams and del Carmen Jorge y Jorge (2008).

[15]See Gandz (1929), who distinguishes a geometry of lines and a geometry of angles.

[16]For a discussion of possible roles of these texts in the surveyors' tradition, see Damerow (2001, 271). A brief outline of the development of Sumerian and Babylonian mathematical practices in their institutional contexts is presented by Høyrup (1994, 4–9).

[17]Robson (2008, 75–83).

cedures of calculation to them. Thus, the relation of lengths to areas, which had earlier been given through fixed assignments,[18] could now be viewed as given by the multiplication of two numbers.[19]

Notwithstanding this formation of more general and abstract arithmetical structures, the 'formulas' for calculating the area of irregular quadrangles (and triangles) remained the ones inherited from the context of surveying, whose results deviate from those of Euclidean geometry whenever the angles involved are not exactly right. That the measure of area defined by this practice was not understood as an approximation of a 'true' (Euclidean) measure becomes clear from the fact that it could even play a constructive role in the understanding of problems. There are, for instance, problems where an area has to be divided into a given proportion, and the length of the dividing line determined. This problem is not well-defined from a Euclidean perspective, because the proportion into which an area is to be divided generally does not determine the length of the dividing line. Nevertheless, if area is defined by the surveyors' formula and the additivity of area is assumed, the solution of the problem becomes unique.[20]

Summing up, the practices of the social control of space in ancient Mesopotamia and the development of the symbolic representations related to them imply several spatial structures which may be characterized as metrical.

- *Conservation of length, area, and volume*: The quantities of length, area, and volume are independent from place or position.
- *Additivity of lengths and distances*: The practice of using standardized measuring rods or ropes and, in particular, their successive apposition, implies that lengths can be added. The practices of numerical notation of lengths and distances reflect this property in the form of algorithms for symbolic addition. The same applies to areas and volumes.
- *Unity of space on different scales*: The integration of units of length, area, and volume into metric systems spanning spaces of different scales (documented on proto-cuneiform tablets from the late fourth millennium) unifies these spaces by making them commensurable.
- *Mutual dependence of different dimensionalities through the quantitative relation of length, area, and volume measures*: Quantitative measures of length, area, and volume make it possible to establish a connection between the measures of different dimensions. Thus, measures of length may be coordinated with measures of area by conventionally fixed assignments (as is documented for early Mesopotamian surveyors). With the possibility of coordinating unit systems for length and area arithmetically, i.e. by means of a general operation of multiplication (as given

---

[18]Schooltexts from 2700 BCE onwards document the need to learn the relation between unit areas and combinations of unit lengths; see the discussion by Damerow (forthcoming, Sect. 3.3).

[19]Owing to the lack of a cipher zero and of a separation between whole units and fractions, the use of the positional system implied the difficulty of keeping track of the order of magnitude for reconversion into the traditional units.

[20]See the discussion of tablet YBC 4675 by Damerow (2001, 280–286).

within the sexagesimal place value system of Babylonian mathematics) their relation becomes even more systematic. The same applies to volume.[21]

The Character of Spatial Knowledge

The spatial knowledge discussed in this chapter ranges from practical knowledge to mathematical knowledge. It is the expert knowledge of a particular group of administrators and develops over history along with the means of symbolic representation. It is externally represented by measurement devices, drawings, and symbolic notation, which develops into writing on one hand and numerical notation on the other.[22] It thereby reproduces structures found on a more elementary level of cognition, this time, however, endowing spatial entities with arithmetic properties. This arithmetization of spatial entities also leads to an integration of spatial structures which, on a more elementary level, remain separated.

The conservation of the size and shape of an object independent of its location and position, for instance, is implied by the sensorimotor schemata responsible for the coordination of perspectives. It is further implicit in the direct comparison of the size of objects when no standardized means of measurement are available. The assumption of the conservation of the size of an object when it is moved through space is, in fact, a precondition for the use of measuring rods or ropes. In the context of the use of such tools and in the presence of standard measures of length, area, and volume, the conservation of size becomes manifest on the level of mathematical representation and implies metric homogeneity of space. The three-dimensionality of objects and spaces is another example. It is perceptually given on the sensorimotor level. Through the mutual (arithmetical) dependence of length, area, and volume it is reflected on the level of the symbolic means of knowledge representation.

Through the application of the sexagesimal place value system, with its general procedures for addition, subtraction, multiplication, and division, and in combination with an abstract system of units defined by its internal relations, the metric structure of space becomes more general and more unified. This illustrates how in certain historical situations the emergence of new means of knowledge representation in specialized practical contexts (surveying) may lead to a dynamic of knowledge development that brings about knowledge structures no longer directly related to that context (Babylonian geometry). But the Babylonian case also shows that this greater generality implicit in the symbolic means of knowledge representation need not become explicit, for instance, in the form of a term that represents the concept of a three-dimensional metric space spanning various scales.

---

[21] Volumes were usually measured in area units, assuming them to be of a conventionally fixed thickness equal to a unit measure of length. If necessary, the thickness was 'raised', i.e., the area was multiplied to obtain a volume of different thickness (Høyrup 2002, 22, 36). The results were sometimes converted to measures of capacity; see Friberg (2007, 196–198).

[22] See Damerow (2012).

Despite its novel degree of abstraction and its thorough metrization of area, the space of Babylonian geometry actually differs from Euclidean space. The procedures of Babylonian geometry are of a limited generality which testifies to their origin in administrative practices. In particular, there is the absence of the consideration of angles as "objects of mensuration,"[23] which is rooted in the implicit definition of area by means of the surveyors' formula.

References

Ascher, M. (1986). Mathematical ideas of the Incas. In M. P. Closs (Ed.), *Native American mathematics* (pp. 261–289). Austin, TX: University of Texas Press.

Chemla, K., & Guo, S. (Eds.). (2004). *Les neuf chapitres: Le classique mathématique de la Chine anciennne et ses commentaires*. Paris: Dunod.

Closs, M. P. (Ed.). (1986). *Native American mathematics*. Austin, TX: University of Texas Press.

Clagett, M. (1992–1999). *Ancient Egyptian science*. Philadelphia, PA: American Philosophical Society.

Damerow, P. (2001). Kannten die Babylonier den Satz des Pythagoras? Epistemologische Anmerkungen zur Natur der Babylonischen Mathematik. In J. Høyrup & P. Damerow (Eds.), *Changing Views on Ancient Near Eastern Mathematics, BBVO 19* (pp. 219–310). Berlin: Reimer.

Damerow, P. (2012). The origins of writing and arithmetic. In J. Renn (Ed.), *The globalization of knowledge in history* (pp. 153–173). Berlin: Edition Open Access.

Damerow, P. (forthcoming). The impact of notation systems: From the practical knowledge of surveyors to Babylonian geometry. In M. Schemmel (Ed.), *Spatial thinking and external representation: Towards an historical epistemology of space*. Berlin: Edition Open Access.

Damerow, P., & Lefèvre, W. (1996). Tools of science. In P. Damerow, *Abstraction and representation. Essays on the cultural evolution of thinking* (pp. 395–404). Dordrecht: Kluwer.

Folkerts, M. (1992). Mathematische Probleme im Corpus agrimensorum. In O. Behrends & L. C. Colognesi (Eds.), *Die römische Feldmeßkunst: Interdisziplinäre Beiträge zu ihrer Bedeutung für die Zivilisationsgeschichte Roms* (pp. 311–336). Göttingen: Vandenhoeck & Ruprecht.

Friberg, J. (2007). *Amazing traces of a Babylonian origin in Greek mathematics*. New Jersey: World Scientific.

Gandz, S. (1929). The origin of angle-geometry. *ISIS, 12*(1929), 452–481.

Guo, S. (Ed.). (1993). *Zhongguo ke xue ji shu dian ji tong hui: Shu xue juan yi*. Zhengzhou: Henan jiaoyu chubanshe.

Høyrup, J. (1994). *In measure, number, and weight: Studies in mathematics and culture*. Albany, NY: State University of New York Press.

Høyrup, J. (2002). *Length, width, surfaces: A portrait of old Babylonian mathematics and its kins*. New York: Springer.

Hyman, M. D. (2006). Of glyphs and glottography. *Language and Communication, 26*(3–4), 231–249.

Koch, G. (1984). *Maligdam. Ethnographische Notizen über einen Siedlungsbereich im oberen Eipomek-Tal, zentrales Bergland von Irian Jaya, West-Neuguinea, Indonesien*. Mensch, Kultur und Umwelt im Zentralen Bergland von West-Neuguinea (No. 15). Berlin: Reimer.

Koch, G., & Schiefenhövel, W. (2009). Eipo (West-Neuguinea, Zentrales Hochland)—Neubau des sakralen Männerhauses in Munggona. DVD, Produktionsjahr: 1974, first published in 1979; IWF Bestellnummer/Bandzählung E 2475.

Lévi-Strauss, C. (1955). *Tristes tropiques*. Paris: Plon.

---

[23]Gandz (1929, 453).

Little John, J. (1963). Temne space. *Anthropological Quarterly, 36*(1), 1–17.

Lyons, H. (1927). Ancient surveying instruments. *The Geographical Journal, 69*(2), 132–139.

Michel, T. (1983). *Interdependenz von Wirtschaft und Umwelt in der Eipo-Kultur von Moknerkon. Bedingungen für Produktion und Reproduktion bei einer Dorfschaft im zentralen Bergland von Irian Jaya, West-Neuguinea, Indonesien.* Mensch, Kultur und Umwelt im zentralen Bergland von West-Neuguinea (No. 11). Berlin: Reimer

Neugebauer, O. (1934). *Vorlesungen über Geschichte der antiken mathematischen Wissenschaften, Vol 1: Vorgriechische Mathematik.* Berlin: Springer.

Renn, J., & Valleriani, M. (2014). Elemente einer Wissensgeschichte der Architektur. In J. Renn, W. Osthues, & H. Schlimme (Eds.), *Wissensgeschichte der Architektur, Vol. 1: Vom Neolithikum bis zum Alten Orient* (pp. 7–53). Berlin: Edition Open Access.

Robson, E. (2008). *Mathematics in ancient Iraq. A social history.* Princeton, NJ: Princeton University Press.

Selin, H. (Ed.). (1997). *Encyclopaedia of the history of science, technology and medicine in nonwestern cultures.* Dordrecht: Kluwer.

Shen, K., Crossley, J. N., & Lun, A. W. C. (Eds.). (1999). *The nine chapters on the mathematical art: Companion and commentary.* Oxford: Oxford University Press.

Thiering, M., & Schiefenhövel, W. (forthcoming). Spatial concepts in non-literate societies: Language and practice in Eipo and Dene Chipewyan. In M. Schemmel (Ed.), *Spatial thinking and external representation: Towards an historical epistemology of space.* Berlin: Edition Open Access.

Vinette, F. (1986). In search of Mesoamerican geometry. In M. P. Closs (Ed.), *Native American mathematics* (pp. 387–407). Austin, TX: University of Texas Press.

Vogel, K. (Ed.). (1968). *Neun Bücher arithmetischer Technik: Ein chinesisches Rechenbuch für den praktischen Gebrauch aus der frühen Hanzeit (202 v. Chr. bis 9 n. Chr.).* Vieweg, Braunschweig

Williams, B.J., & del Carmen Jorge y Jorge, M. (2008). Aztec arithmetic revisited: Land-area algorithms and Acolhua congruence arithmetic. *Science, 320*(72), 13–27.

Chapter 5
Reflection and the Context-Independence of Mental Models of Space

Abstract From ancient societies such as Greece and China there is evidence for the-oretical reflections on the representations of elementary and practical forms of spatial knowledge. In both societies this development can be argued to be closely related to the emergence of cultures of disputation and a vivid tradition of writing. The chapter discusses the knowledge emerging from such reflections as being distinguished from the elementary and practical forms on which it builds by its greater generality and aspiration for consistency. Two traditions are taken as examples, both originating in ancient Greece and both being further pursued, under different social and cultural circumstances, up into modern times: (1) the tradition of deductive geometry, which originated in the reflection on practical knowledge involving the use of drawing instruments; and (2) the tradition of philosophies of space, which originated in the reflection on the linguistic representation of elementary spatial knowledge.

Keywords Reflection · Euclidean geometry · Non-Euclidean geometry · Philoso-phy of space · Aristotelianism · Atomism

The Object of Study

The cultural developments of spatial thinking discussed in the preceding chapter show a basic trend to cognitive structures that are less dependent on the specific practical contexts from which they originated. An example is the emerging practice of area determination by means of a multiplication of lengths within the sexagesi-mal system, which implies a greater independence of the concept of area from the context of surveying than any conventional way of relating areas to standard lengths based on specific practices of measurement and notation. The increase in generality is obviously related to the development of the means of knowledge representation such as comprehensive systems of units and a place-value number system. But this development is only the material side of a dialectical process whose other side is mental. When performing operations on external knowledge representations, cogni-tive structures are built up which are mental reflections of these operations. Since these operations disregard many aspects of the real-world objects, this mental process may be referred to as a *reflective abstraction*. When the new mental structures are in turn externally represented, e.g., by symbols forming a system, we may speak of a representation of higher order than the one the process of reflection started from.[1]

Processes of reflective abstraction are a consequence of the exploration of existing means of knowledge representation. Exploration of these means by individuals may happen spontaneously at any time in history. But such individual developments remain without consequences in the history of knowledge unless there are social entities such as organized groups or institutions that ensure that the cognitive products are handed down and—at least for some period—become subject to cumulative development. We already encountered a candidate case of such an institutionalization: the schools of the scribes in Mesopotamia which brought about Babylonian geometry as a doctrine of areas independent of the context of surveying—even though the structure of Babylonian geometry still bears witness to its origin in practical surveying (see the previous chapter). The context of teaching and learning to handle the symbolic means of knowledge representation seems to be a natural place for the emergence of explorative forms of knowledge. Another such context is disputation, traditions of controversial discourse and rational debate. While such traditions are usually oral in origin, they may find expression in text traditions, possibly accompanied by an ongoing oral component. Disputation is a motor for reflection on concepts and, as a consequence, for their generalization. The resolution of apparent paradoxes, for instance, presupposes reflection on language and the delineation of meanings. Spatial knowledge need not be the primary object of these reflections, but when comprehensiveness is aspired to, it will naturally come into consideration.

One may distinguish two types of explorative knowledge, which may roughly be designated *mathematical* and *philosophical*. Mathematical explorative knowledge results from systematic reflection, specifically upon representations related to the use of instruments such as measuring rods and ropes, the straight edge, and the compass.[2] Philosophical explorative knowledge, by contrast, results primarily from systematic reflection upon the linguistic representations of elementary shared knowledge.

Most prominent among the historical settings in which the exploration of the cognitive tools of spatial thinking became productive are the intellectual traditions of ancient Greece. The first-order knowledge that was reflected upon in this context was by no means of purely Greek origin. From the Archaic period on, astronomical, medical, and arithmetic knowledge from Egypt and Mesopotamia entered the Greek world.[3] In contrast to the Babylonian case, which was defined by the needs of central state administrations, the Greek situation was characterized by polycentrism, the encounter of different strata of society, and the negotiation and public justification of political decisions.[4] Upon this background systematic reflections were pursued which aimed at establishing a coherent, encompassing world view, distinct from the

---

[1] See Damerow (1996, 371–381).

[2] On the role of language as a means of knowledge representation in the emergence of theoretical mathematics, see Lefèvre (1981).

[3] See Schiefsky (2012) for a concise discussion and references to the literature.

[4] Lefèvre (1981), Lefèvre (1984, 306), Hyman and Renn (2012, 86–87).

received mythology but with the same aspiration for totality. Written texts produced in the context of the Greek philosophers' activities now provide us with the earliest evidence of systematic reflections on the linguistic representation of shared spatial knowledge. A parallel and related development is the formation of a characteristic Greek tradition of mathematics, particularly concerned with questions of geometry.[5]

The later historical intellectual places which furthered a deliberate and purposeful exploration of the implications of systems of knowledge representation included: the Neoplatonic schools of late antiquity; hellenistic science pursued at places like the Museion of Alexandria; the court science, philosophy and theology of the Arab Middle Ages pursued in places like Bagdad and Córdoba; and the scholasticism of the Latin Middle Ages. In early modern times the theoretical reflection on fundamental concepts such as space and matter gained a new impetus in the context of an ideological struggle between different strata of society. In their attempts to formulate encompassing counter world systems against the predominantly Aristotelian world view promoted by the Church, early modern natural philosophers faced the challenge of taking into account an increasing amount of empirical knowledge from practical mathematics and astronomy. In the following centuries, theoretical reflection on space has become increasingly institutionalized in the disciplinary discourses of physics and philosophy.

While these historical places of systematic reflection are all related more or less strongly by ties of tradition—all of them are rooted in one way or another in the theoretical traditions of Greek antiquity—there is at least one example of an independent emergence of systematic reflection on spatial language. The so-called *Mohist Canon* from the Warring States Period in China (ca. 300 BCE) documents such an independent tradition and thereby represents a rare source for addressing comparative questions in the long-term history of spatial knowledge; questions concerning the conditions for the emergence of traditions of systematic reflection and the necessities and contingencies in their development. The analysis of passages in the *Mohist Canon* and their comparison to Western sources have shown, for instance, that the occurrence of elementary mental models in theoretical thinking on space is indeed a cross-cultural phenomenon. The connection of such reflections with encompassing natural philosophies, by contrast, is a peculiarity of the Greek case and depends on the timing of specific theoretical traditions such as the construction of cosmologies on one hand and the reflection on the meaning of words on the other.[6]

When the context-independence of mental models of space resulting from reflection is investigated, the objects of study must be the traditions of explorative knowledge documented in texts from the different historical periods and places described above. In the following, two traditions shall be sketched and some of their implications on basic structures of spatial thinking shall be outlined: (1) the mathematical

---

[5]On the institutional background of the emergence of Greek mathematics, see Høyrup (1994, 9–15), who explicitly contrasts the Greek with the Babylonian case and argues for a close connection of the emergence of Greek mathematics with the contemporary philosophical discourse.

[6]See Boltz and Schemmel (2015). For a comprehensive account on the Later Mohist's logic, ethics, and science, see Graham (1978).

tradition of deductive geometry associated with the name of Euclid; and (2) the philosophical tradition of encompassing theories of place and space originating in Greek antiquity.

Example: Deductive Geometry

The Babylonian involvement with the geometry of fields entailed the construction of plans (see the foregoing chapter). There were other practical activities in ancient civilizations, such as house building, in the contexts of which geometrical constructions were produced, even involving plans drawn to scale.[7] All these constructions, whether true to scale or not, constitute representations of relations within empirical space and were most probably produced for planning purposes. The role of geometrical constructions changed fundamentally when they became the object of systematic reflections in Greek mathematics. The tradition that culminated in the composition of Euclid's *Elements* presents us, in that work, with a deductive system of statements about the figures one can draw by means of compass and straight edge. Euclidean geometry thus emerges from the reflection on the products of a graphical practice performed on a limited range of scale and relying on a very restricted set of instruments. At the same time, its application in various practical and scientific contexts such as surveying and astronomy demonstrates that its validity with respect to physical space was generally assumed.

The *Elements* implies a metrization of space more general than that of Babylonian proto-geometry. The concept of length, for instance, is no longer defined by a particular system of units of measure and the corresponding operations of measurement and calculation, but is implicitly defined in the *common notions*, an element of the text's deductive structure, wherein it is stated how magnitudes compare. Thus, the first three common notions read[8]:

1. Things which are equal to the same thing are also equal to one another.
2. If equals be added to equals, the wholes are equal.
3. If equals be subtracted from equals, the remainders are equal.

In this manner, *the additivity of length, area, and volume* implied by the spatial practices discussed in the preceding chapter is made explicit in Euclidean geometry. The other metrical structures implied in spatial practices, such as the *conservation of length, area, and volume*, the *unity of space on different scales*, and the *mutual dependence of different dimensionalities* equally hold in Euclidean geometry and are partly made explicit. Thus, the third postulate, which states that it is possible "[t]o

---

[7] See Heinrich and Seidl (1967) and Heisel (1993) on ancient Near Eastern ground plans; the latter work also discusses Egyptian, Greek, and Roman plans.

[8] Euclid (1956, I, 155).

describe a circle with any centre and distance,"[9] presupposes the *unity of space on different scales*.

The Euclidean concepts are also more general in that they take angles fully into account. Thus, the major line of argument which may be discerned in Book I of the *Elements* leads up to a proposition that makes it possible to reduce any rectilineal figure to a rectangle of equal area (proposition I, 45).[10] This construction stands in stark contrast to the Mesopotamian methods of area determination, which do not take angles into account, and therefore implies a different concept of area. Similarly, the concept of length, or distance between two points, is enriched by the network of propositions. In particular, it is endowed with a metric structure equally taking angles into account. One way to see this makes use of the Pythagorean theorem (proposition I, 47), which states that the square of the hypothenuse of a rectangular triangle equals the sum of the squares of its legs. Employing analytic geometry (which only developed in early modern times[11]), we can use the theorem to define the linear distance Δs of any two points: its square is the sum of the squared differences in their Cartesian coordinates.

$$(\Delta s)^2 = (\Delta x_1)^2 + (\Delta x_2)^2 \tag{5.1}$$

The metric distance is the length of the shortest path between two points. Therefore, when applied to the space of one's motion, it gives a quantitative meaning to the elementary spatial knowledge about the *dependency of the effort on the path taken* when getting from one place to another. In doing so, it is disregarding all non-geometric factors that could make other paths preferable, e.g., slopes, obstacles, conditions of the ground, or, when seafaring is concerned, winds and currents. (Further, distances have to be small enough so that the sphericity of the earth can be neglected.)

Among the forms of knowledge representation involved in the Euclidean *Elements*, we may distinguish two orders.[12] The geometrical figures the text reflects upon may be called *first-order representations*, since they directly present what can be drawn with straight edge and compass. The deductive structure of the text, by contrast, is part of the linguistic *second-order representation*, which is a representation not of real-world objects, but of mental objects, viz., ideal mathematical objects and the metric structure they imply. The genetic relation between the two orders of representation is reflected in various aspects of the way the text of the *Elements* presents geometrical knowledge. In particular, within the deductive structure, two kinds of propositions are distinguished: theorems, coming along with a proof of their validity, and construction tasks. The presence of the construction tasks, which come along

---

[9]Euclid (1956, I, 154).

[10]Mueller (2006, 16).

[11]The publication in 1637 of Descartes' *Geometry* (Descartes 1925) was a milestone in the development of analytic geometry. Coordinate systems with rectilinear axes at right angles to each other, commonly referred to as *Cartesian*, are a later development, however.

[12]For this and the following, see Damerow (1994, 268–270, 277).

with a proof that the constructed figures possess the required properties, are a clear indication that the text results from the reflection on constructed figures as first-order representations of spatial relations, and that such figures played a constitutive role in its emergence.

The second-order representation of geometrical knowledge in a deductive system played a double, on the face of it paradoxical, role in the history of spatial thinking. On the one hand, it consolidated the appearance of Euclidean geometry as something that reflected necessities of human spatial cognition (irrespective of the question if these were considered to result from properties of the outside world or from universal preconditions of human cognition). Being logically derived from a set of axioms that state obvious truths, the theorems of geometry could hardly be called into question.[13] On the other hand, it was the very reflection on such second-order representations that eventually led to a generalization of geometrical concepts beyond Euclidean geometry. This process is most prominently illustrated by the century-long unsuccessful attempts to prove the dependence of the fifth postulate on the other postulates.[14] The postulate, which is also known as the *postulate of parallels*, states that, if a straight line cuts two other straight lines in such a way that the sum of the interior angles on one side of the cutting line is less than two right angles, the two lines, when infinitely produced, will meet on that side. The self-evidence of the postulate was questioned early on, for instance by Proclus.[15] The attempts to derive it led to various reformulations. Over the course of the nineteenth century its independence from the other postulates became evident when it was discovered that consistent geometries can be constructed which do not obey the postulate of parallels.[16]

The development of these non-Euclidean geometries generalized geometrical concepts in such a way that Euclidean geometry appears as nothing but a special case in a continuum of possible geometries.[17] When only geometries of constant curvature are considered, this curvature may be positive, negative, or zero, in which case we speak of spherical (elliptic), hyperbolic, and Euclidean space, respectively. While basic structures such as *the conservation* and *the additivity of length, area, and volume, the unity of space on different scales*, and *the mutual dependence of different dimensionalities* still hold in these non-Euclidean geometries, the distance between two points in three-dimensional space is no longer generally given by Eq. (5.1). As a

---

[13]This does not imply, however, that the Euclidean axioms were always regarded as representing objective truths. Salomon Maimon, for instance, held the view that while the Euclidean deductions from axioms represent objective truths, since they are based on the understanding, the synthetic axioms are valid only subjectively, since they depend on human intuition; see Freudenthal (2012, 137–138).

[14]For a brief survey of such attempts, see Heath's note on the fifth postulate (Euclid 1956, I, 202–220).

[15]See Proclus' commentary on the fifth Postulate of Euclid's *Elements*, Book I; for an English translation, see Proclus (1970, 150–151).

[16]For an outline of the long-term transformation of the object of geometry from figures to second-order properties of figures, and eventually to space, which was a precondition for the formulation of non-Euclidean geometries, see De Risi (2015, 1–13).

[17]Klein (1968, 188–211).

consequence, certain properties of figures which are independent of size in Euclidean geometry become dependent on it. The sum of the interior angles of a triangle, for instance, always equals two right angles in Euclidean geometry. In elliptic geometry it is greater, in hyperbolic geometry it is less, the deviation from two right angles increasing in both geometries in proportion with the area of the triangle (and in both geometries there is an upper limit to the possible size of a triangle). But there are even more general types of geometries when the requirement of constant curvature is dropped. Requiring space to be locally Euclidean, i.e., that at any point in space, coordinates can be found so that an infinitesimal version of Eq. (5.1) applies, the infinitesimal distance between two points in two-dimensional space becomes

$$ds^2 = g_{11}dx_1{}^2 + 2g_{12}dx_1dx_2 + g_{22}dx_2{}^2, \qquad (5.2)$$

where the g_{ij} are functions of the coordinates x_i, $i = 1, 2$.

The discovery of the possibility of non-Euclidean geometries generalized the concept of geometry and, at the same time, destroyed the apparent necessity in the relation between Euclidean geometry and physical space. On the background of the common conception of Euclidean geometry as a theory of physical space, even on cosmological scales, the development of non-Euclidean geometries naturally brought up the question of their applicability to this space. The pioneers of non-Euclidean geometries themselves, such as Carl Friedrich Gauss, Nikolai Lobachevsky, and Bernhard Riemann, all speculated about such an applicability.[18]

The reflection on the epistemic status of the axioms of geometry led to a new awareness of the importance of things in space, in particular rigid bodies and light rays,[19] for the determination of spatial geometry. Thus, Hermann von Helmholtz, following Kant halfway, distinguished a transcendental part of spatial cognition, our a priori intuition of space, which included the three-dimensionality and the continuity of space, and an empirical part which presents us with rigid bodies that can be used for measuring space, so that the question of which geometry (of constant curvature) applies to physical space becomes an empirical one[20]:

> [...] if to the geometrical axioms we add propositions relating to the mechanical properties of natural bodies, were it only the axiom of inertia, or the single proposition, that the mechanical and physical properties of bodies and their mutual reactions are, other circumstances remaining the same, independent of place, such a system of propositions has a real import which can be confirmed or refuted by experience, but just for the same reason can also be gained by experience.

Note how Helmholtz, on a theoretical level of argument, reintroduces practical structures of cognition such as the *conservation of length*, thus reminding us of the empirical origins of geometry. The insight that the metrical structure of space remains

---

[18] See Torretti (1978, 61–67 on Gauss and Lobachevsky and 103–107 on Riemann).

[19] On the fundamental role of light rays for the concept of a straight line, see Eisenstaedt (2012).

[20] Helmholtz (1962, 245).

undetermined as long as the physical content of space is not taken into consideration is a strong argument against a priori-conceptions of geometry.[21]

The observation that the geometry of physical space can only be investigated with the help of physical bodies was also the basis for Henri Poincaré's conventionalism, according to which the question of the geometry of physical space was merely a matter of convenience; since the geometry of space cannot be determined independently from a theory of physical bodies, the former could always be kept Euclidean by appropriately changing the latter.[22]

Reflections on the second-order representation of geometrical knowledge eventually revealed that this knowledge could be represented completely independently from spatial intuition, in fact without any reference to space whatsoever. David Hilbert's work on the foundations of geometry[23] presents a fully axiomatic theory, in which Euclidean geometry is reduced to a system of sentences, some of them representing axioms and all others being derivable from them. Points, lines, and surfaces are implicitly defined by their mutual relations; an intuition about what they are in space is unnecessary for the construction of the theory. Although the role of intuition for the justification of the axioms remained a controversial issue, the relation of mathematical theory to physical space had changed. On the one hand, there was the purely mathematical structure, regardless of its origins, while on the other there was the physical application of this structure of which it was unknown to which degree of accuracy it worked.

Starting from such considerations, Albert Einstein concluded that while Poincaré was right "[s]ub specie aeterni"[24] when arguing for his conventionalism, the present state of physics did not allow one to get rid of measuring rods and clocks, i.e., rigid bodies, in order to link geometry to physical space. Einstein's argument may be understood as a reference to a hierarchical structure of knowledge in which different layers interact. Since the conceptual structure of physical theories (special and general relativity) remains related to more fundamental spatial experiences (measuring distances and time intervals with rods and clocks), the new experiences captured by these theories may affect our understanding of space and time. It was in fact the reflection on this relation between the operations of measurement and the axioms of geometry that allowed Einstein to consider non-Euclidean geometries as describing physical space and time (see Chap. 7).[25]

As we have seen in the context of the practice of surveying, the conservation of the size of objects is a precondition of the concept of a metric space. Any use of measures is meaningful only if there is a property of bodies, called size, which does

---

[21] Torretti (1978, 170).

[22] Poincaré (1902). On conventionalism, and specifically on Poincaré's geometrical conventionalism, see Ben-Menahem (2006).

[23] Hilbert (1903).

[24] Einstein (1921, 8). Einstein formulates his ideas on the relation between axiomatic and practical geometry in his essay *Geometrie und Erfahrung* (Einstein 1921); an English translation is found in Einstein (2001, 208–222).

[25] Einstein (1921, 6–7).

not change under transformations such as rotation or translation in space. In the context of reflections on geometry, the first-order representations of spatial knowledge, like measuring rods, attain a new meaning. They now become higher-order representations with an important role in theories of space. They ensure the possibility of connecting mathematical theories to physical space.

Example: Theories of Space

Long before attaining any abstract or technical meaning, spatial language reflects aspects of the sensorimotor knowledge structures described in Chap. 2. Words like *empty* or *void*, for instance, play a role in the linguistic representation of the *dichotomy of objects and spaces*, according to which there are bodies and empty spaces between them. The use of a word like *empty*, for instance when referring to an empty vessel, further bears the potential for its generalization: everything that is not a body may be conceived of as an *empty space*. Yet, linguistic representations do not spontaneously give rise to general concepts such as that of an *empty space*. They attain such general meanings only when they become part of a conceptual system, or theory. This has prominently happened in ancient Greek atomism, which provides one of the earliest documented cases of a systematic reflection on elementary knowledge structures concerning space and matter.

The atomists argued that the world is composed of just two things: the atoms (ἄτομα) and the void (κενόν). The atoms were conceived of as extended but nevertheless indestructible particles which moved through the void. The atoms in their extendedness, solidity, and movability are clearly modeled after the permanent objects experienced in the living environment. The void, on the other hand, is modeled after the empty spaces between them. In the theories of the atomists, the *dichotomy of objects and empty spaces* thus becomes absolute. The concepts no longer pertain directly to aspects of the living environment. While an object in the living environment may be hard, it is never indivisible in principle. And while the empty space between two objects may not be tangible, it usually contains air and may be thought of as being filled with various other things. The empty space of ancient atomism, by contrast, is absolutely empty, it is even referred to as *nothing*.[26]

In these theories, the *void* is not only the empty space between the atoms, but also the space in which they reside. This becomes clear when motion is considered. Just as empty space is a collection of potential places for atoms, the actual places of atoms are filled 'empty spaces'. This view is clearly expressed by Lucretius who in his *De rerum natura* states that[27]

---

[26]See Simplicius *In Aristotelis De caelo commentaria*, 294, 33–295, 24; Diels (1951–1952, II, 93–94(68 A 37)); for a German translation, see Jürß et al (1988, 118–119). The identification of 'the empty' with 'nothing' points to the discursive context in which atomism was brought forward, namely the refutation of the Parmenidean idea of one only absolute and static being. As a reaction, the atomists postulated the existence of the non-existing, i.e., of the *nothing*.

[27]Lucretius, *De rerum natura*, I, 419–421; translation from Lucretius (1992, 35–37).

[...] the nature of the universe, therefore as it is in itself, is made up of two things; for there are bodies, and there is void, in which these bodies are and through which they move this way and that.

The atomistic concept of the void is no longer restricted to concrete contexts of action involving such things as empty vessels. It is supposed to be a term describing a trait of the entire world, a trait that lies at the foundation of all appearances whatsoever. It is therefore the earliest attested concept of space that abstracts from the objects in space. It may be referred to as the *container model* of space. Empty space is a container for the atoms and thereby for everything in the world. Although the atomists' description of the void as *nothing* is purely negative, it must evidently be conceived of as possessing some of the spatial properties we discerned in the case of sensorimotor mental models of space and objects, such as *three-dimensionality* and *path-connectedness*. If the direction of fall of heavy bodies is fixed prior to any particular configuration of bodies, as is evidently assumed by Lucretius, the *distinction of the vertical direction* is a further property of space rather than a property of the bodies it contains.[28] This implies that space, while being homogeneous, is anisotropic.[29]

Aristotle not only reflects upon the linguistic representation of spatial knowledge, but makes such a reflection his explicit program when he says that[30]

The Natural Philosopher has to ask the same question about 'place' as about the 'unlimited'; namely, whether such a thing exists at all, and (if so) after what fashion it exists, and how we are to define it.

The generalizations that follow from Aristotle's reflections on spatial concepts are quite different from those of the atomists. Aristotle does not take the *dichotomy of objects and spaces* as an absolute foundation. Yet, in his theoretical thinking we can equally identify aspects of space implied by the sensorimotor mental models of object and space. Aristotle's considerations build upon the *definiteness and exclusivity of place*, the idea that every object is in a place and that two objects cannot simultaneously be in the same place. Accordingly, the central term in Aristotle's discussion is *place* (τόπος) rather than *space* (χώρα), as the above quote indicates.[31] The place of a body is defined as the inner surface of the body surrounding the said body. It is thus "a vessel that cannot be moved."[32] Accordingly, Aristotle denies the

---

[28]Lucretius uses the distinction of the vertical direction in the context of an argument for the infinity of space, claiming that if space were finite, all bodies would have collected in a heap on the bottom of the universe; Lucretius *De rerum natura*, I, 984–997, see Lucretius (1992, 83).

[29]See Jammer (1954, 11).

[30]Aristotle *Physics*, IV, 208a 27–29, cited after Aristotle (1993, 277).

[31]Aristotle mentions *space* (χώρα) only five times in his *Physics* (208b7, 209a8, 209b12, 13, 15), in the first two instances using expressions such as "place and space" and in the remaining instances relating to Plato. A recent discussion of the relation of τόπος and χώρα in Aristotle's *Physics* is found in Fritsche (2006a) and Fritsche (2006b). For a discussion of the development of Aristotle's concept of *place*, also taking into account the *Categories*, see, e.g., Mendell (1987). For a monograph on Aristotle's concept of place, see Zekl (1990).

[32]Aristotle *Physics*, IV, 212a 15–16.

existence of a void. Place is always the place of a body with respect to other bodies, so that without body there is no place. Rather than absolutizing the dichotomy of tangible objects and empty spaces between them, Aristotle thus builds upon the experience that everywhere there is something. Between the tangible objects there is air which can be perceived, for instance, in the case of wind.

As is the case for the atomists, Aristotle's reflections on the linguistic representation of elementary spatial knowledge lead to general spatial concepts that are not restricted to any particular context of action. But Aristotle's concept of place does not equally lead to a conception of space as existing independent of bodies as is the case for the atomists' void. Contrasting Aristotle's conception with the atomistic container model of space, it can be described as conceiving of space as the "positional quality of the world of material objects."[33] While Aristotelian *place* may be defined for very large collections of contiguous bodies, the cosmos as a whole has no place because there is no body surrounding it. Accordingly, there is no place or space. Yet the Aristotelian concept of place has a cosmological dimension. Places qualify direction, and they do so differently for different bodies. Heavy bodies being primarily composed of earth and water naturally move towards the geometrical center of the spherical cosmos as the natural place of earth; light bodies being primarily composed of fire and air naturally move towards the periphery, their natural place being spheres just below the lunar sphere, which is the innermost of the celestial spheres. This anisotropic organisation of the cosmos can be understood as a reflection on the elementary *distinction of vertical direction*, combined with the insight into the sphericity of the earth which was part of the astronomical knowledge of the time.[34]

When elementary structures of spatial cognition become the foundation for theories of space, as happens with atomistic and Aristotelian physics, the implications of these structures are explored outside their original realm of validity. This exploration forces the theoreticians to decide between alternatives and to argue for these decisions. In the realm of elementary knowledge, there is, for instance, no contradiction between the *dichotomy of objects and spaces* on one hand and the notion that everywhere there is something on the other. Both ideas reflect experience and apply in their respective contexts of action. But when these ideas are elevated to the rank of general principle, the consequent exploration of their implications is bound to lead to contradictions. How these contradictions are to be resolved is not determined by the structures of elementary knowledge. This indeterminateness is a general aspect of the systematic reflection upon the linguistic representations of elementary spatial knowledge. Structures of spatial thinking unquestioned in the realm of elementary knowledge may become contested in the realm of the philosophy of space.

---

[33] Einstein in his foreword to Max Jammer's *Concepts of Space* (Jammer 1954, xi–xvi); the quote is on p. xiv. In his foreword, Einstein introduces and discusses the fundamental distinction between the concepts of space as the container for all things and space as the positional quality of all things.

[34] This insight was ignored by Lucretius as his argument for the infinity of space mentioned above demonstrates (see note 28). At the same time it was widely known to astronomers and geographers (see Chap. 6). This fact sheds light on the uneven spread of astronomical and geographical knowledge amongst different intellectual groups within ancient Roman society.

This does not mean that 'anything goes' in the philosophy of space. The reflective handling of linguistically represented knowledge presupposes intersubjectively shared standards of argumentation. Theories of space can be motivated and arguments for and against them can be weighed in the context of more encompassing systems of knowledge. Criteria such as consistency, generality, empirical adequacy, and, in corresponding historical and cultural contexts, theological adequacy may be invoked.[35] But the inherent indeterminacy of theoretical generalizations from elementary knowledge structures explains the persistent occurrence of controversies about the fundamental properties of space. These controversies were not a peculiarity of ancient Greek philosophy but also subsisted through late antiquity, the Arab and Latin Middle Ages, and well into modern times. Does a void exist? Is space finite or infinite? Is space continuous? With respect to what is motion to be attributed to a body? These are examples for fundamental questions about space that could not be answered conclusively in the framework of philosophical theorizing. In the different historical contexts of discussion, we see similar argumentative constellations reoccur, which derive from the rootedness of theories of space in elementary and practical knowledge structures. Let us sketch some of these argumentative constellations for the four questions just formulated.

Does a void exist? If the elementary *dichotomy of objects and spaces* is absolutized, as in ancient atomism, the answer is *yes*. The answer becomes less straightforward, however, even in the context of the container model of space, when the dichotomy of objects and spaces is modified by the introduction of third entities, such as an imponderable substance filling all of space, e.g., an ether or light.[36] If the term 'void' is considered to designate the absence of ponderable matter (for instance on the background of the imponderable substance being identified with space itself) a void may still exist.[37] If, by contrast, the imponderable substance filling all of space is considered some kind of body, there can be no absolute void even in the context of the container model.[38]

---

[35] In fact, the aspiration to argumentative justification even affects the discussion about elementary structures that are uncontroversial. The *three-dimensionality of objects and spaces*, for instance, describes an aspect of space consistent with both conceptions of space, the container model and the position-quality model, and still became the object of argumentative justifications; see, for example, Aristotle's line of reasoning in *De caelo*, 268a6–b5 (Aristotle 1986, 4–7).

[36] On the relation between light and space in Proclus, Robert Grosseteste, and others, see, for instance, Jammer (1954, 36–40). A more detailed discussion of Proclus is given in Sambursky (1977, 180–182), which deals with the concepts of place and space in late Neoplatonism (see also Sambursky (1982), which contains the related sources with translations).

[37] This appears to be the standpoint of Giordano Bruno when in *De l'infinito, universo et mondi* he lets Filoteo say: "We do not call aught Void as being mere nullity, but rather accept the view whereby that which is not corporeal nor doth offer sensible resistance is wont, if it hath dimension, to be named Void [...]" (Singer 1968, 273); for the original Italian and a German translation, see Bruno (2007, 98–99).

[38] This is the gist of an argument by Gottfried Wilhelm Leibniz (who himself is not a proponent of the container model of space) against the alleged proof of the vacuum in experiments such as Otto von Guericke's. In his fifth letter to Samuel Clarke he writes: "The Aristotelians and Cartesians, who do not admit a true vacuum, have said in answer to that experiment of Mr. Guerike, as well as

The most consequent conceptual elaboration based on the position-quality model of space is the identification of material and spatial extension.[39] In such a framework a void is utterly inconceivable because without a body there is no extension. Accordingly, it can be argued that a truly empty vessel's walls would touch since between them there would be nothing.[40] An alternative way of conceiving of space as nothing but an aspect of the world of bodies, without conflating the concepts of space and body, is the Aristotelian conception according to which the spatial concept of place is not identified with three-dimensional extension but with the two-dimensional inner surface of the containing body. The concept of place is thus distinguished from the extension of the body under consideration, but is at the same time inseparably linked to the surrounding bodies. In such a framework, the existence of a void may again be refuted by pointing out that extension does not exist independent of body, since place is a relation between bodies and not an extension which may be conceived devoid of body.[41] This refutation of the existence of void thus hinges on the relational definition of place. Deviating from this definition by conceiving place as a three-dimensional immaterial extension, distinguished from the material extension of bodies, as Philoponus did in his commentary on Aristotle's *Physics*, means to abolish this kind of refutation; the resulting conception of place is based on the container model of space rather than the position-quality model. Even in this model it remains possible, of course, to argue for the absence of a void within the cosmos.[42] The arguments cannot be as fundamental as those based on the position-quality model, however, since the theoretical possibility of a vacuum can no longer be excluded.

Given that a primary intuitive meaning of the concept of existence is informed by the elementary model of a permanent object which implies *tangibility* and *being in a certain place at a time*, it becomes understandable that, in the context of theoretical reflection, the ontological status of empty space presents a difficulty.[43] Therefore, empty space was often not given the ontological status of a substance, even by those who argued for its existence, but rather of an attribute. But it had to be the attribute

---

(Footnote 38 continued)

to that of Torricellius of Florence, [...] that there is no vacuum at all in the receiver; since glass has small pores, which the beams of light, the effluvia of the load-stone, and other very thin fluids may go through. I am of their opinion: and I think the receiver may be compared to a box full of holes in the water, having fish or other gross bodies shut up in it; which being taken out, their place would nevertheless be filled up with water." (Alexander 1970, 65).

[39] A prominent proponent of this view was Descartes who held that space is nothing more and nothing less than what he considered to be the defining property of bodies, namely extension; *Principles of Philosophy*, Part 2, in particular §§ 4, 9, 10, 11, 12 (Descartes 1984, 40–41,43–45). There were medieval and early modern predecessors holding such a view, among them John Buridan, Franciscus Toletus, and Francisco Suarez; see Grant (1981, 14–17) on 'internal space'.

[40] Descartes, *Principles of Philosophy*, Part 2, § 18 (Descartes 1984, 47–48).

[41] Aristotle *Physics* IV, 7, 214a, 16–19.

[42] On Philoponus, see Grant (1981, 19–21).

[43] This problem of the existence of a philosophical void was perceived from the very beginning when the early atomists claimed 'the existence of the non-existing'. In the sequel, historical discussions about the existence of a void often touched upon the question of what it means for something to *be* (Cf. Grant 1981, 9–23).

of something *being*, so that, in empty space, something had to *be* that was not filling space but whose attribute space was.[44]

Besides these fundamental arguments about the existence of a void, which are based on elementary cognitive structures constituted by the mental models of objects and spaces, arguments for or against the void may be put forward that require further theoretical presuppositions and physical observations. Although in many cases these presuppositions, too, can be shown to be rooted in elementary knowledge structures, the related arguments are less cogent and are accordingly taken to be less essential by many historical commentators. An example is Aristotle's arguments against the existence of a void based on a discussion of the possibility of motion in a hypothetical void. In one of these arguments, Aristotle rejects the void by arguing that motion in it would be instantaneous. In the Middle Ages, these arguments were controversially discussed and the idea of instantaneous motion in a void was widely refuted. At the same time, Aristotle's rejection of the void was generally accepted, a fact that suggests that the arguments relating to motion in a void were not regarded as being decisive.[45] Atomists, by contrast, argued that a void was necessary to allow for motion. The early Greek atomists had introduced the void in particular because motion in a plenum was considered impossible. But the argument presupposes a certain conception of matter and can be refuted, for instance, by referring to objects floating in water.

Is space finite or infinite? While experiential space is always finite, because all spatial experience is limited, it is possible in the imagination to go beyond any given boundary. This is the basis for Archytas' classic argument for the infinity of space, in which it is imagined that an arrow be shot beyond the alleged boundary of space.[46] The argument clearly works within the container model of space, according to which a material boundary of that container is an impossibility, because it must itself be within the container-space. Accordingly, most theories of space based on the container model assume space to be infinite. This holds for the ancient as well as for the early modern atomists, including Isaac Newton. Within the position-quality model, on the other hand, space is just an aspect of the material cosmos. If the material cosmos is finite, there is no foundation for the existence of space beyond it. In Aristotelian-Ptolemaic cosmologies, in which the most distant objects, the fixed stars, make a full revolution about the center of the universe in one day, the spherical shell in which they reside cannot be infinitely extended; usually it is assumed to be a rather thin shell. In Copernican-type cosmologies, in which the earth rotates and the stars remain fixed, by contrast, the infinite extension of the realm of fixed stars becomes a possibility again.[47]

---

[44] An example of such a line of argument is again found in the Leibniz-Clarke correspondence (Alexander 1970, 37, 47, and 66–72, no. 8 in Leibniz' fourth letter, Clarke's reply to it, and Leibniz' reply to Clarke's reply).

[45] For the medieval discussions on motion in void space, see Grant (1981, 24–60).

[46] See Sorabji (1988, 125). The argument is reiterated by Lucretius, *De rerum natura*, I, 968–983 (Lucretius 1992, 81–82).

[47] The historical transition *from the closed world to the infinite universe* is famously discussed by Alexandre Koyré (1958). For a more recent account, see Omodeo (2014, 158–196).

The dilemma that arises from the possibility, within metaphysics, to argue for and against both assumptions, finiteness and infinity of the cosmos, is the subject of Kant's first antinomy of pure reason.[48] When non-Euclidean geometries are taken into account, new possibilities to approach the question of infinity arise. In particular, it becomes conceivable that space is finite but unbounded, just like the surface of a sphere.[49] To such a universe Archytas-type arguments do not apply. But in order to eliminate them fully it is necessary to conceive of three-dimensional space as being curved without being embedded in a higher-dimensional flat space. While such a concept of space is perfectly possible mathematically, the intuitive analogy with curved surfaces breaks down at this point, since the latter can only be imagined embedded within three-dimensional space. The case of non-Euclidean cosmologies therefore provides an example of how the development of mathematical formalisms leads to physical concepts drastically modifying the set of theoretical options grounded in the mental models of elementary cognition.[50]

Is space continuous? Taking the elementary *continuity of object trajectories* as the point of departure, the intuitive answer is *yes*. Yet, taking basic mathematical knowledge about lengths into account, the continuity of space becomes problematical. In particular, the *additivity of lengths* may be invoked to decompose a line into ever smaller parts. Continuity then means that this decomposition may be performed *ad infinitum*. The ultimate constituents of a line then have to be of such sort that by adding an infinite number of them one would obtain a finite line segment.[51] But this seems to lead into a dilemma: either these constituents are of finite size, then an infinite number of them will yield a line of infinite length; or they are dimensionless, then any number of them (including infinity) will remain dimensionless.[52] One way out of this dilemma is to deny continuity and assume that there are 'atoms' of geometry, smallest units of space of finite size. When space is conceived as merely

---

[48] Kant (1996, 458–464).

[49] Edgar Wind made explicit use of this hypothesis to resolve the first Kantian antinomy, Wind (2001a, 131–159); for an English translation, see Wind (2001b). Karl Schwarzschild, long before the rise of relativistic cosmology, tried to estimate the largest possible curvature of such a universe (Schwarzschild 1900; see the discussion in Schemmel 2005).

[50] The role of formalisms in the transformation of mental models is further discussed in the following chapter.

[51] It is not necessary to imagine the concept of such constituents as arising from some kind of idle contemplation. More probably it emerged in the context of a sophisticated mathematical tradition, e.g., when it may have been hoped that it was useful in the attempt to unite incommensurable magnitudes by providing a basic unit for both the side of a square and the square's diagonal; see the discussion by Boyer (1959, 20–21).

[52] This is one of the famous paradoxes of Zeno of Elea. An elementary discussion of different paradoxes attributed to Zeno and their resolution by means of modern mathematical concepts is given in Huggett (1999, 37–50). The paradox could be resolved only when it was discovered that the number of points in any finite (or infinitely large) line segment is *uncountably infinite* rather than countably infinite. While a countably infinite number of dimensionless points does not constitute a line of finite length, an uncountably infinite number does.

an aspect of extended bodies, or of relations between them, the idea of discontinuity can be motivated by reference to smallest, indivisible units of bodies (atoms).[53]

With respect to what is motion to be attributed to a body? The elementary *dichotomy of movable and unmovable objects* implies that motion is always conceived of as relative to something considered at rest. What is considered at rest and what in motion is determined by the practical context. In the large-scale space of motion the landmarks are usually considered at rest, but we have also seen a case in which the role of the moving and the unmoving were reversed (the example of Carolinian navigation discussed in Chap. 3). When motion and rest become the object of reflection, their context-dependence may explicitly be noted. A man on a boat may be at rest relative to the boat and may, at the same time, be moving with the boat relative to the shore. But as long as the environment with its landmarks—trees, mountains, buildings—is considered at rest, it can always be argued that, for any object, there is only one 'true' or 'absolute' motion (including rest), namely its motion with respect to this environment. In cosmologies in which the earth is considered at rest, this concept of absolute motion can be transferred to all motion in the universe, including celestial motions and the motion of hypothetical entities such as atoms. In cosmologies based on inhomogeneous or anisotropic spaces, motion can further be related directly to the spatial structures, rather than to actual bodies such as the earth, whose position may itself be explained as contingent on the structure of space. Thus, in Aristotelian cosmology, it is the geometric center of the spherical cosmos that defines the directions of the natural motions and the center of the circular celestial motions, not the body of the earth. It is therefore geometrically defined places that function as landmarks. But while the Aristotelian cosmos is inhomogeneous and anisotropic, it is spherically symmetric, so that circular motions around its center cannot be described with respect to some abstract structure. The Lucretian cosmos, by contrast, is homogeneous and anisotropic, so that no other aspect of motion but the direction of the motion of fall is prescribed by the structure of space.

If the earth is not considered at absolute rest, one can argue in different ways. One can argue that there is no such thing as absolute motion and that all motion is relative. One can also argue that there are still certain objects serving as landmarks for describing motion, but that they are not landmarks of global validity, but only with respect to the body under consideration. As a consequence, the only 'true' motion of a body is its motion with respect to its immediate surrounding bodies.[54] If, by contrast, the sun and the fixed stars are considered at absolute rest, as is the

---

[53] Such an argument appears to have been made for time and motion and, possibly, also for space by the Mutakallamin (Islamic dialecticians); see Gent (1971, 46–47), Jammer (1954, 60–66), and Gosztonyi (1976, 150–154). The idea of indivisible lines as minimum spaces appears to have also been entertained by Francesco Patrizi; see Henry (2001).

[54] Both arguments are given by Descartes in his *Principia*, Book 2, §§ 24, 25 (Descartes 1984, 50–51). Both (mutually incompatible) concepts of motion do not allow for a definition of what distinguishes uniform motion in a straight line, a concept that Descartes needs in his pioneering statement of the law of inertia. Descartes' definition of motion "properly speaking" is sometimes interpreted as resulting from an attempt to avoid persecution by the anti-Copernican Church; see, e.g., Barbour (1989, 442–444).

case for the Copernican view of the cosmos, then absolute motion is described with respect to them. Any point in the cosmos is then absolutely fixed by its distance from the center of the sun and the direction, given with respect to the stars, of the line joining the center of the sun and the point. If, however, no celestial body is known to be at absolute rest, the only material body left to relate absolute motion to is a hypothetical motionless ethereal substance filling all of space.[55] But motion may also be related to non-material entities such as space itself. When absolute motion is motion relative to an absolute space, space must provide the 'landmarks' with respect to which motion is described. We have seen that the structures of inhomogeneous and anisotropic spaces may provide such landmarks. But what if space is conceived of as perfectly homogeneous and isotropic? In this case the points and regions that make up space must be individuated prior to the consideration of any material contents of space, so that it makes sense to describe motion relative to them. One may think of a rigid coordinate-system fixed at absolute rest which would provide a set of sublimate landmarks.[56]

Besides the landmark model, there is another elementary cognitive structure related to motion that can be exploited in theoretical discussions about true or absolute motion. This is the mental model according to which *motion implies force*.[57] This mental structure reflects the elementary experience that one has to exert a force in order to move something. In the context of theoretical reflections on relative and absolute motions, the model may be employed by conceiving as absolute only those motions that are effected by a force.[58] In Newtonian physics, the motion-implies-force model is modified (see the following chapter). Force is now proportional to acceleration, i.e., to the change of the state of motion, not to the presence of motion. The inertial forces occurring in a system of reference are then interpreted as evidence for an accelerated motion of the system with respect to absolute space. In Newtonian physics, the effects of accelerated motion thus function as indicators of a

---

[55] This line of reasoning is found in Schwarzschild (2007, 185–186). Schwarzschild refers to the so-called proper motions of the stars which were noticed towards the end of the seventeenth century when observation techniques were further refined. He goes on to remark that the electromagnetic ether is probably not without motion. (On the electromagnetic ether and its state of motion, see the following chapter.)

[56] The idea that one can individuate the points of space prior to the consideration of bodies was severely criticized by Leibniz in his correspondence with Clarke on the basis of the principle of the identity of indiscernibles; see Alexander (1970, 26 and 36–39).

[57] See Renn and Damerow (2007) for a description of this model and its historical transformations.

[58] This is indeed Leibniz's criterion to distinguish absolute from relative motions, see, e.g., paragraph 53 in his fifth paper of his correspondence with Clarke: "[...] I grant there is a difference between an absolute true motion of a body, and a mere relative change of its situation with respect to another body. For when the immediate cause of the change is in the body, that body is truly in motion [...]" (Alexander 1970, 74). At the same time, Leibniz refused to relate the concept of absolute motion to a concept of absolute space. It seems that he saw no necessity of postulating a universal reference frame to relate all absolute motions to. This implies that the relative velocity of two bodies in absolute motion is not given by the difference of their absolute velocities. In other words, if we know the absolute motion of two different bodies, we still do not know their motions relative to one another.

spatial structure that provides landmarks (for instance by defining directions). This argument combines the modified motion-implies-force model with the landmark model of space. The two models do not readily match, however, since the landmark model defines absolute rest, while the modified motion-implies-force model does not distinguish rest from uniform motion in a straight line.

The Character of Spatial Knowledge

The spatial knowledge discussed in this chapter can be described as *theoretical knowledge*. This kind of knowledge is to a great extent conditioned by its means, i.e., by the external knowledge representations from the exploration of which it emerges. It is handed down in text traditions, mostly in the form of written language and symbolic notation, which make it possible to pick up a tradition even centuries after it has last been actively pursued. It is aimed at consistency and comprehensiveness and thereby gives rise to more general and abstract concepts such as those of Euclidean distance and absolute void, in many cases including a general concept of space.

The explorative reflection upon elementary structures of spatial thinking brings about theoretical structures which preserve many of the spatial properties implied by sensorimotor intelligence. At the same time, the theoretical context of generalization and the aspiration for consistency leads to questions about these properties which never could have occurred in elementary or practical contexts. At the level of fully developed sensorimotor activity, the mental models have their clear-cut realm of application. At the level of theoretical thinking, by contrast, there is an inherent uncertainty about which aspects of the mental models to build upon. This ambiguity derives from the absence of the concrete contexts of action that limit the meaning of the linguistic representations of knowledge in their everyday use. The operations of external representations in reflective thinking are dissociated from these original contexts and bring out structures inherent in the system of representations. The result of such processes of reflective abstraction are not predetermined in general, because the space of possible structures spanned by the means of representation is much richer than any particular realization in it.

There is a striking difference between philosophical and mathematical explorative knowledge. While the former turned out to depend on individual decisions (considered within systems that integrated a larger range of knowledge), and remained controversial throughout the history of philosophical thinking, the latter was, from early on, considered to present inevitable truths. The well-defined object of reflection of mathematical explorative knowledge, the first-order representations of instrumental knowledge, allowed for a consistent representation within a deductive structure. Nevertheless, the further reflection upon the higher-order representations of Euclidean geometry eventually led to theoretical alternatives (the validity of non-Euclidean geometries) that could not be evaluated on purely rational grounds, recalling the case of philosophical explorative knowledge.

The reflection on first-order representations (constructed figures) led to a generalization of spatial concepts which implied a de-contextualization: what was a theory of constructed figures became interpreted as a theory of space, decoupled from what fills space. The reflection on second-order representations (deductively organized sets of statements) then further generalized the spatial concepts, but at the same time brought about the re-contextualization of geometry when the role of rigid bodies and light rays for establishing the geometry of physical space was appreciated. The emergence of non-Euclidean geometries thus functioned as a historical reminder of the empirical origins of Euclidean geometry in instrumental action. Accordingly, and in spite of deviating epistemological claims, the question of the applicability of non-Euclidean geometries turned out to be an empirical question. In this context, first-order representations of spatial knowledge (measuring rods), became higher-order representations, which relate the abstract structures to physical space by relating theoretical knowledge to other layers of knowledge.

References

Alexander, H. G. (Ed.). (1970). *The Leibniz-Clarke correspondence together with extracts from Newton's 'Principia' and 'Optics'*. New York: Manchester University Press.

Aristotle, (1986). *On the heavens*. Aristotle in twenty-three volumes. Cambridge, MA: Loeb Classical Library, Harvard University Press.

Aristotle, (1993). *The physics*. Aristotle in twenty-three volumes. Cambridge, MA: Loeb Classical Library, Harvard University Press.

Barbour, J. B. (1989). *Absolute or relative motion? Vol. 1: The discovery of dynamics*. Cambridge, MA: Cambridge University Press.

Ben-Menahem, Y. (2006). *Conventionalism*. Cambridge, MA: Cambridge University Press.

Boltz, W. G., & Schemmel, M. (2015). Theoretical reflections on elementary actions and instrumental practices: The example of the 'Mohist Canon'. Preprint 468, Max Planck Institute for the History of Science, Berlin; to appear in: M. Schemmel (Ed.), *Spatial thinking and external representation: Towards an historical epistemology of space*. Berlin: Edition Open Access.

Boyer, C. B. (1959). *The history of the calculus and its conceptual development (the concepts of the calculus)*. New York: Dover.

Bruno, G. (2007). *De l'infinito, universo et mondi: Italienisch - Deutsch*. Hamburg: Meiner.

Damerow, P. (1994). Vorüberlegungen zu einer historischen Epistemologie der Zahlbegriffsentwicklung. In G. Dux & U. Wenzel (Eds.), *Der Prozeß der Geistesgeschichte: Studien zur ontogenetischen und historischen Entwicklung des Geistes* (pp. 248–322). Frankfurt: Suhrkamp.

Damerow, P. (1996). *Abstraction and representation: Essays on the cultural evolution of thinking*. Dordrecht: Kluwer.

De Risi, V. (Ed.). (2015). *Mathematizing space: The objects of geometry from antiquity to the early modern age*. Cham: Springer.

Descartes, R. (1925). *The geometry of René Descartes: With a facsimile of the first edition 1637*. Chicago, IL: The Open Court Publishing Company.

Descartes, R. (1984). *Principles of philosophy*. Dordrecht: Reidel.

Diels, H. (1951–1952). *Die Fragmente der Vorsokratiker* (6th ed.). Berlin: Weidmann, revised by W. Kranz.

Einstein, A. (1921). *Geometrie und Erfahrung*. Berlin: Springer.

Einstein, A. (2001). *The collected papers of Albert Einstein, English translation of selected texts* (Vol. 7). Princeton, NJ: Princeton University Press.

Eisenstaedt, J. (2012). La petite histoire de la ligne droite qui se mord la queue. *Quadrature, 91,* 1–6.

Euclid, (1956). In Heath, T. L. (Ed.), *The thirteen books of Euclid's Elements* (2nd ed.). New York: Dover.

Freudenthal, G. (2012). Anschauung und Verstand in geometrischen Konstruktionen: Kant und Maimon. In S. Dietzsch & U. Tietz (Eds.), *Transzendentalphilosophie und die Kultur der Gegenwart* (pp. 113–138). Leipzig: Leipziger Universitätsverlag.

Fritsche, J. (2006a). Aristotle on 'chora' in Plato's 'Timaeus'. *Archiv für Begriffsgeschichte, 48,* 27–44.

Fritsche, J. (2006b). Aristotle on space, form, and matter. *Archiv für Begriffsgeschichte, 48,* 45–64.

Gent, W. (1971). *Die Philosophie des Raumes und der Zeit. Historische, kritische und analytische Untersuchungen.* Hildesheim/New York: Georg Olms.

Gosztonyi, A. (1976). *Der Raum. Geschichte seiner Probleme in Philosophie und Wissenschaften.* Freiburg: Alber.

Graham, A. C. (1978). *Later Mohist logic, ethics and science.* Hong Kong: Chinese University Press.

Grant, E. (1981). *Much Ado about nothing: Theories of space and vacuum from the Middle Ages to the Scientific Revolution.* Cambridge, MA: Cambridge University Press.

Heinrich, E., & Seidl, U. (1967). Grundrißzeichnungen aus dem Alten Orient. *Mitteilungen der Deutschen Orient-Gesellschaft zu Berlin, 98,* 24–45.

Heisel, J. P. (1993). *Antike Bauzeichnungen.* Darmstadt: Wissenschaftliche Buchgesellschaft.

Henry, J. (2001). Void space, mathematical realism, and Francesco Patrizi Da Cherso's use of atomistic arguments. In C. H. Lüthy, J. E. Murdoch, & W. R. Newman (Eds.), *Late medieval and early modern corpuscular matter theories* (pp. 133–161). Leiden: Brill.

Hilbert, D. (1903). *Grundlagen der Geometrie* (2nd ed.). Leipzig: Teubner.

Høyrup, J. (1994). *In measure, number, and weight: Studies in mathematics and culture.* Albany, NY: State University of New York Press.

Huggett, N. (Ed.). (1999). *Space from Zeno to Einstein: Classic readings with a contemporary commentary.* Cambridge, MA: MIT Press.

Hyman, M., & Renn, J. (2012). Survey: From technology transfer to the origins of science. In J. Renn (Ed.), *The globalization of knowledge in history* (pp. 75–104). Berlin: Edition Open Access.

Jammer, M. (1954). *Concepts of space: The history of theories of space in physics.* Cambridge, MA: Harvard University Press.

Jürß, F., Müller, R., Schmidt, E. G., et al. (Eds.). (1988). *Griechische Atomisten: Texte und Kommentare zum materialistischen Denken der Antike* (3rd ed.). Leipzig: Reclam.

Kant, I. (1996). *Critique of pure reason: Unified edition; with all variants from the 1781 and 1787 editions.* Indianapolis, IN: Hackett.

Klein, F. (1968). *Vorlesungen über nicht-euklidische Geometrie, 1928 edition.* Berlin: Springer.

Koyré, A. (1958). *From the closed world to the infinite universe.* New York: Harper.

Lefèvre, W. (1981). Rechensteine und Sprache. In P. Damerow & W. Lefèvre (Eds.), *Rechenstein, Experiment, Sprache: Historische Fallstudien zur Entstehung der exakten Wissenschaften* (pp. 115–169). Stuttgart: Klett-Cotta.

Lefèvre, W. (1984). Die Wissenschaft in der geschichtlichen Entwicklung des Menschen. In N. Loacker (Ed.), *Kindlers Enzyklopädie: Der Mensch* (Vol. 7, pp. 295–328). Zurich: Kindler.

Lucretius, (1992). *De rerum natura.* Cambridge, MA: Loeb Classical Library, Harvard University Press.

Mendell, H. (1987). Topoi on topos: The development of Aristotle's concept of place. *Phronesis, 32,* 206–231.

Mueller, I. (2006). *Philosophy of mathematics and deductive structure in Euclid's Elements.* Mineola, NY: Dover.

Omodeo, P. D. (2014). *Copernicus in the cultural debates of the Renaissance: Reception, legacy, transformation.* Leiden: Brill.

Poincaré, H. (1902). *La Science et l'hypothèse*. Paris: E. Flammarion.

Proclus, (1970). In Morrow, G. R. (Ed.), *A commentary on the first book of Euclid's Elements*. Princeton, NJ: Princeton University Press.

Renn, J., & Damerow, P. (2007). Mentale Modelle als kognitive Instrumente der Transformation von technischem Wissen. In H. Böhme, C. Rapp, & W. Rösler (Eds.), *Übersetzung und Transformation* (pp. 311–331). Berlin: de Gruyter.

Sambursky, S. (1977). Place and space in late Neoplatonism. *Studies in History and Philosophy of Science, 8*(3), 173–187.

Sambursky, S. (Ed.). (1982). *The concept of place in late Neoplatonism*. Jerusalem: The Israel Academy of Sciences and Humanities.

Schemmel, M. (2005). An astronomical road to general relativity: The continuity between classical and relativistic cosmology in the work of Karl Schwarzschild. *Science in Context, 18*(3), 451–478.

Schiefsky, M. (2012). The creation of second-order knowledge in ancient Greek science as a process in the globalization of knowledge. In J. Renn (Ed.), *The globalization of knowledge in history* (pp. 191–202). Berlin: Edition Open Access.

Schwarzschild, K. (1900). Über das zulässige Krümmungsmaass des Raumes. *Vierteljahrsschrift der Astronomischen Gesellschaft, 35*, 337–347.

Schwarzschild, K. (2007). Things at rest in the universe. In J. Renn & M. Schemmel (Eds.), *Gravitation in the twilight of classical physics: Between mechanics, field theory, and astronomy* (pp. 183–190). Dordrecht: Springer.

Singer, D. W. (1968). *Giordano Bruno: His life and thought. With annotated translation of his work: On the infinite universe and worlds*. New York: Greenwood Press.

Sorabji, R. (1988). *Matter, space and motion: Theories in antiquity and their sequel*. Ithaca, NY: Cornell University Press.

Torretti, R. (1978). *Philosophy of geometry from Riemann to Poincaré*. Dordrecht: Reidel.

Wind, E. (2001a). *Das Experiment und die Metaphysik: Zur Auflösung der kosmologischen Antinomien*. Frankfurt: Suhrkamp.

Wind, E. (2001b). *Experiment and metaphysics: Towards a resolution of the cosmological antinomies*. Oxford: European Humanities Research Centre.

Zekl, H. G. (1990). *Topos: Die aristotelische Lehre vom Raum*. Hamburg: Meiner.

Chapter 6
The Expansion of Experiential Spaces Over History

Abstract Besides the reflection on representations of existing spatial knowledge, the expansion of spaces of experience is a motor for conceptual development, whether these are the geographical spaces known through political expansion, trade, and exploration, the cosmological spaces known through observation, or the microcosmic spaces known through engineering and experimentation. The chapter presents three examples for processes of concept formation and the generalization of spatial concepts that were promoted by such expansions of experiential spaces. The first example refers to the systematic accumulation of geographical knowledge, which laid the foundation for the introduction of a global system of terrestrial coordinates. This allowed landmarks to be related no longer just to other landmarks but also to a mathematically determined, abstract geographical space. The second example refers to the accumulation over centuries of astronomical and mechanical knowledge, which, by a process of reflective integration, brought about the Newtonian concept of a homogeneous, isotropic, absolute space independent of its matter content. The third example relates to the expansion of knowledge of microscopic space by institutionalized research on electric and magnetic forces, which brought about and stabilized the concept of the electromagnetic field.

Keywords Experiential knowledge · Institutions · Geographical coordinates · Absolute space · Field concept · Formalism

The Object of Study

In the previous chapter we have seen that reflection on the external representations of elementary and practical knowledge may lead to new and more general spatial concepts. In such cases of theoretical thinking, novelty arises from the structures inherent in the means of knowledge representation and tools for intellectual labor becoming explicit through their exploration and through reflective abstraction. But the history of theoretical reflection does not unfold before a background of unchanging spatial experience. When interested in the relation of experience and theoretical reflection in the historical development of spatial concepts, one has to take into account a

© The Author(s) 2016
M. Schemmel, *Historical Epistemology of Space*, SpringerBriefs
in History of Science and Technology, DOI 10.1007/978-3-319-25241-4_6

complementary long-term trend: the expansion of experiential spaces. In this chapter it is argued that this expansion of experience not only implies an accumulation of spatial knowledge but rather plays an important role in bringing about new spatial concepts and stabilizing them within more encompassing knowledge systems.

Starting with the first steps of ontogenesis, experience plays an instrumental role in shaping human spatial cognition (Chap. 2). Beyond the immediate experiential environment of the individual, different socially shared spaces can be experienced in different societies. This experiential basis of spatial knowledge expanded in the course of history, not monotonically and not universally, but under a long-term, global perspective. One may distinguish three realms of experiential space to which this expansion pertains. First of all it pertains quite literally to the *geographic spaces* known to human societies, which have grown through travel, trade, exploration, and military campaigns. Such activities brought about the expansion of the space of movement of various societies or even of their organized space, as is the case with expanding empires which bring increasingly more land under political and economic control. These spaces have grown in many local historical contexts and also in a long-term perspective, spanning the time from prehistoric nomadic and sedentary tribes to modern global societies enabling intercontinental travel and communication.

Another experiential space that has expanded over history is *cosmological space*. Cosmological space is the entire universe known, or assumed to exist, by a given society. To this space society transfers spatial concepts and knowledge acquired in terrestrial contexts. It is in particular also the space of mythological realms of experience. Cosmological space is experiential through the observation of the sky, in particular systematic astronomical observation. This space has vastly grown from observations of the sun, the moon, the planets, and the stars in early societies, to the modern observation of astronomical objects billions of light years away. It has also grown with respect to the richness of its physical contents. With the increasing refinement of celestial mechanics from antiquity to modern times, and with the rise of astrophysics in the course of the nineteenth and twentieth centuries—developments clearly related to the progress of observational instruments and techniques—the import of knowledge from terrestrial science into cosmology has vastly increased. With the observation of the flight of the galaxies, cosmological space itself has been turned into an object onto which elements of physical description such as the field equations of general relativity or the model of a black body may be applied. Visible light has become just one within a wide range of sources of knowledge about the universe and present-day astronomy reaches the brink of what is physically observable: looking far away means looking back in time and with the most recent developments in the research on the detection of gravitational waves there is the justified expectation that one will soon be able to 'look through' the early universe, which is opaque with respect to electromagnetic radiation.

Microcosmic space, just like macrocosmic space, has been a target for the projection of experiential knowledge from the mesocosmic realm, as the example of atomism discussed in the previous chapter illustrates. On the background of such theoretical world views, knowledge about physical objects acquired through practical experiences in dealing with technological artifacts or even through systematic experimentation potentially has implications for spatial concepts. The further expansion of experiential knowledge about the micro-world was not only due to new instruments of magnification—from the optical microscope to the particle-accelerator—but also to the systematic exploration of chemical, electric, and magnetic phenomena. In particular the increase, in modern times, of empirical knowledge in the fields of mechanics and electrodynamics brought about fundamental changes of the concept of space, as shall be demonstrated below.

When considering the impact of the expansion of experiential spaces on spatial thinking, the object of study is the processes of concept formation fostered by the increase of experiential knowledge in the three realms described above: geographical, cosmological, and microcosmic space. In this chapter, examples from all these realms shall be discussed. From these examples it will also become clear that the conceptual developments in the three realms are closely intertwined. The three examples are:

- *the development of global coordinates for geographical space* in late hellenistic times as a consequence of the Greco-Roman expansion;
- *the independence of space from matter and force* through the establishment of the concept of absolute space in early modern physical science as a consequence of the integration of empirical knowledge about terrestrial and celestial motions; and
- *force fields as a third entity between objects and space* emerging as a new concept in nineteenth century physics as a consequence of the expansion of electrical and magnetic experiences.

Example: The Development of Global Coordinates for Geographical Space

Ancient Greece and Rome experienced expansions of known geographic space hardly matched by any earlier civilization. From archaic times on, and particularly in the period from roughly 750 to 550 BCE, Greek civilization expanded, scattering into hundreds of colonies spread over an increasingly large portion of the Mediterranean and Black Sea Regions.[1] In hellenistic times, Alexander's campaigns opened up territories as far to the East as the Indus River. Increased travel and trade activities were partly facilitated by the discovery of Monsun winds. The Roman Empire expanded into new territories, from the British Islands and Germania in the north, to the Kingdom of Kush in modern day Sudan in the south. Trade relations included India and China.

The expansion of known, or even controlled, spaces went along with an increasing institutionalization of the acquisition and processing of geographical knowledge. The acquisition of geographical knowledge was among the deliberate aims of Alexander's campaigns, which involved experts of different specialized professions such as surveyors who produced itineraries during their travels. At the hellenistic

---

[1] Malkin (2011).

royal courts and the Museion of Alexandria, the integration of new geographical information with the existing geographical knowledge was systematically pursued.[2] For the Romans, the structuring of space, for instance by building streets with milestones indicating distances, was an explicit means of securing control. Roman and Parthian campaigns served as a basis for further geographical knowledge, as was famously presented by Strabo in his *Geography*.[3]

The Greco-Roman expansion of experiential space led not only to the accumulation of an increasingly large corpus of geographic knowledge but to qualitative changes in the way this knowledge was structured and externally represented. Of crucial importance in this context is the parallel development of celestial knowledge and its cosmological implications. When one observes the sky, exactly what one sees depends on geographical location. This has to be taken into account when astronomical instruments or instruments employing astronomical occurrences, such as the sundial, are carried from one place to another. This means, in turn, that the differences in the celestial occurrences can be used to distinguish geographical places. This relation has played an increasingly important role in the history of Greek cartography from early on up to its culmination in Ptolemy's use of a global system of geographic coordinates.

This development implies certain cosmological assumptions. Most importantly it implies the hypothesis of the spherical shape of the earth. Other assumptions relate to the sphericity of the sky, the centrality of the earth within the celestial sphere, and the smallness of the terrestrial sphere when compared to the celestial sphere.[4] It is plausible to assume that the observation of differences in the celestial occurrences as they depend on geographical location, such as differences in gnomon shadow lengths at noon, played a role in the emergence of the hypothesis of a spherical earth located at the center of a spherical universe. But whatever the actual historical relation between experience and theoretical thinking at the origin of this hypothesis, the accumulated geo-astronomical knowledge clearly stabilized the hypothesis, which became a basis for the integration of an ever larger amount of empirical knowledge about the inhabited world.[5]

Although the surviving sources of Greek cartography are fragmentary, one can discern certain developments. In particular, increasing importance is attached to the

---

[2]Harley and Woodward (1987, 148–149).

[3]See Strabo (1982–1995) for a modern edition with an English translation.

[4]On the different arguments for the centrality of the earth found in Aristotle and Ptolemy, see Omodeo and Tupikova (forthcoming).

[5]The hypothesis of a close relation between astronomical instruments and cosmological theory is discussed in Szabó and Maula (1982). That the knowledge about the dependency of the celestial appearances on geographical position does not necessarily lead to the hypothesis of a spherical earth is suggested by the Chinese case. The very same phenomena that Eratosthenes exploited in his estimation of the size of the spherical earth (see below)—the variation along the north-south direction of the gnomon shadow length at noon—were used in China to determine the height of the heavens above a flat earth. (I am grateful to Irina Tupikova for pointing this out to me.) See Cullen (1976, 122–127).

dependency of observable celestial occurrences on geographical position.[6] Little is known about early maps, such as that of Anaximander from the first half of the sixth century BCE, which is mentioned by later writers, in particular Strabo. A map is mentioned in Aristophanes' play *The Clouds*, indicating that at his time (around 400 BCE), maps were something commonly known.[7] The relation between the latitude of one's location on earth and celestial observation was early used by the navigator and astronomer Pytheas who flourished around 320 BCE and determined the latitude of Marseilles. In his voyages, Pytheas experienced a wide latitudinal range and may have been among the first to relate latitude to the length of the longest day and draw parallels "to indicate all the places where identical astronomical phenomena could be observed."[8] In the third century BCE, Eratosthenes used the dependence of the gnomon shadow length on latitude to estimate the circumference of the earth. With him we find a question which will occupy cosmographers throughout antiquity and again in early modern times: What is the size of the inhabited land and what is its position on the terrestrial globe? Crates of Mallos, who flourished around 150 BCE, constructed a globe with four separate inhabitable land masses, two on each hemisphere, one of the northern ones being the *oikoumene*. Strabo later also locates the known inhabited world in a northern quadrant of the globe.

Once the relation between celestial and terrestrial positions was established, structures discerned on the celestial sphere could be projected onto the terrestrial sphere. This pertains, first of all, to the different celestial circles that were employed in the description of celestial phenomena. The projection of the parallel circles such as the tropics onto the earth, gave rise to the distinction of five different climate zones on earth. These are, for instance, discussed by Aristotle.[9] An important precondition for the establishment of global terrestrial coordinates was then the development of the designation of stellar positions, which shifted from the description of their place within constellations to the use of coordinates. Hipparchus made use of stellar coordinates such as the distance from the pole in degrees and the distance along the ecliptic from a given constellation. He explicitly emphasized the necessity of using astronomical observations to determine geographical positions ("climata" and distances to the East or the West).[10] Claudius Ptolemy, in his *Syntaxis mathematica* (better known under the Title *Almagest*), which he composed before his *Geography*, lists more than 1,000 stars in ecliptical coordinates.[11]

With Claudius Ptolemy's *Geography*, the grid of celestial coordinates is completely projected onto the earth. In analogy to what he achieved for stellar positions in his *Almagest*, Ptolemy lists the coordinates for thousands of places. He criticizes, but at the same time builds upon, the work of Marinus of Tyre, who seems to have

---

[6]Harley and Woodward (1987, 130–176), Dilke (1987).

[7]See Harley and Woodward (1987, 138–139).

[8]Harley and Woodward (1987, 151).

[9]Aristotle, *Meteorologica* II, v, 362a33–363a20 (Aristotle 1987, 178–185); see also Harley and Woodward (1987, 145).

[10]See Harley and Woodward (1987, 166).

[11]Stückelberger and Graßhoff (2006–2009, 10–11).

grouped places of either the same latitude or the same longitude, but did not systematically provide both coordinates for all the places. Ptolemy integrates Marinus' data
with data from various other sources, including the pole heights of different places
as given by Hipparchus and latest travel reports and maps.[12]

The Ptolemaic projection of celestial structures onto the globe is something quite
distinct from the use of the star compass by the Micronesian navigators, although both
practices share the feature that they make use of celestial knowledge for terrestrial
orientation. The Micronesian navigators used stellar constellations as landmarks for
determining directions within a local environment. Their knowledge of the sky was
useful to them in conjunction with knowledge of the directions to various islands.
In the Ptolemaic case, by contrast, celestial knowledge is used to establish a global
system of geographical positions, completely independently of what actually exists
at the respective locations. In contrast to types of maps that primarily represent topological relations between landmarks such as major paths between named locations,
rivers, mountain ranges, coastal lines, and so on,[13] Ptolemaic maps reflect the attempt
to embed landmarks into a context-independent framework, the framework of global
terrestrial coordinates. The use of such a system of global coordinates implies a
transformation of the mental model of large-scale space according to which space
is structured by a hierarchy of conspicuous, unmovable objects and their relations.
Geographical locations can now be described by reference to the coordinates, which
cover the entire space and function as sublimate landmarks, rather than by reference
to actual landmarks. The landmark model, on which the concrete mental models of
large-scale space are based, is transformed into a model of space spanned by these
sublimate landmarks: 'sublimate' because they are numerically defined rather than
bodily; 'landmarks' because they define fixed positions to which actual landmarks
can be coordinated. The landmark model of space thus becomes more independent
of what fills space; since the coordinates can be defined independently of, and are
prior to, any actual landmark, their use lends autonomy to space with respect to the
bodies filling it and the landmark model becomes assimilated to the container model.

Example: Independence of Space from Matter and Force

In early modern times, gradual but profound changes occurred within the European knowledge system, particularly affecting natural philosophy and, by implica-
tion, the concept of space. An important aspect of these changes, which eventually
brought about modern science, was the integration of different strands of knowledge

---

[12]On Ptolemy's sources, see Stückelberger and Graßhoff (2006–2009, 16–20).

[13]A famous example of such a 'topological' map is the Roman *Peutinger Map* from around 300
CE, which represents the *oikoumene* in a highly distorted way; see, e.g., Talbert (2007).

traditions, which had formerly been separated by social boundaries.[14] Mechanical knowledge, in particular, became increasingly systematized and entered the construction of encompassing world views, a traditional domain of natural philosophy. In this process, two realms of practical mathematics became integrated, both resting on experiential knowledge that had accumulated over millennia, and both witnessing a boost of new empirical knowledge: terrestrial and celestial mechanics.

Practical knowledge of mechanics is closely related to human tool use and long predates any theory of mechanics. It has long been pointed out that, despite the absence of any indication that they were in possession of a science of mechanics, there is ample evidence of mechanical knowledge in the early civilizations of Egypt and Mesopotamia.[15] Our earliest evidence of mechanics as a mathematical science stems from the time of Aristotle.[16] In the ensuing tradition, in which the names of Archimedes, Heron, and Pappus play an outstanding role, mechanical experiential knowledge is presented in the context of the working of simple machines such as the lever and the screw. In its Arabic transmission, mechanical knowledge was focused on the balance and presented in the context of its practical motivations and consequences. It was thereby transformed into a *science of weights* which, in the Latin Middle Ages, was taken up again in an academic context, most prominently by Jordanus de Nemore who flourished in the thirteenth century.[17] Another important medieval development was the doctrine of intension and remission, whose mathematical elaboration flourished in the thirteenth century in Oxford, where the *calculatores* worked at Merton college, and Paris, where Nicolas Oresme devised a method of graphical representation of change.[18] While these schools explicitly considered motion and the change of qualities *secundum imaginationem*, i.e., without relation to observation and experiment, their new conceptual and geometrical tools made it possible to explore the theoretical implications of elementary knowledge on motion far beyond what can be found in Aristotle's work. In early modern times, the static and the kinematic traditions of mechanics were integrated in the works of Galileo Galilei, Thomas Harriot, and other mathematician-philosophers who contributed to what may be called *preclassical mechanics*.[19] Preclassical mechanics developed in the context of a huge expansion of practical knowledge which became an increasingly

---

[14]See Zilsel (2000b), in particular Chap. 2, *The Sociological Roots of Science*, which was first published in 1942. For a broader investigation of the societal causes of the emergence and persistence of modern science, see Lefèvre (1978).

[15]See Mach (1989, 1–3).

[16]The authorship of the *Mechanical Problems* is disputed, which is why its author is usually referred to as Pseudo-Aristotle. There is a contemporary, independent emergence of theoretical reflections on mechanical arrangements and phenomena in China, documented in the *Mohist Canon*, which apparently had virtually no influence on the later course of Chinese intellectual history; see Renn and Schemmel (2006).

[17]Abattouy et al. (2001, 4–5 and 9–10).

[18]See, for instance, Maier (1952) and Clagett (1959). On the use of the diagrammatic representation of motion in early modern science, see Schemmel (2014).

[19]On Galileo, Descartes, and Beeckmann, see Damerow et al. (2004); on Harriot, see Schemmel (2006) and Schemmel (2008).

vital resource of early modern societies. The workings of a variety of mechanical arrangements and phenomena, such as the projectile trajectory, the flywheel, the pendulum, and machines that work by utilizing the force of percussion, presented challenges to mechanical theory. The late medieval and early modern acceleration in the development of mechanical engineering and the technology of warfare not only motivated attempts at further theoretical penetration, but also provided a wealth of new experiential knowledge theory had to account for. Practice even included systematic trials such as the compilation of range tables for the use of gunners, trials which could be turned into scientific experiments with the explicit aim of advancing theory. This expansion of the mesocosmic space of experience and the attempts at taking account of this knowledge by means of mathematically informed theories of nature brought about results that became cornerstones of later classical mechanics, such as the law of fall or the insight into the parabolic shape of projectile trajectories.

The systematic collection of experiential knowledge concerning cosmic space reaches further back in history than that of mesocosmic or microcosmic space. The early civilizations in Mesopotamia, China, and Mesoamerica all developed calendars that temporally structured societal life by referring it to celestial regularities, and engendered institutions devoted to the regular observation of celestial events that were interpreted as omens. In Mesopotamia, such observations are documented starting from the Old Babylonian period, i.e., as early as the first half of the second millennium BCE.[20] The historical development of experiential knowledge concerning cosmic space is special in yet another regard. It is cumulative in a very concrete sense, namely in that, over the course of history, ancient observations are complemented, not replaced, by newer, more precise ones. The old observations remain valuable for all times, because they cannot be repeated and are indispensable for the study of long-term changes. In ancient Greece, the accumulated knowledge about the motions of the fixed stars, the sun, the moon, and the planets became represented in an elaborate geometrical scheme of the cosmos, a development first and foremost associated with the names of Hipparchus and Ptolemy. Ptolemy's geocentric system provided the framework for the accumulation of further observational knowledge in the Arab Middle Ages.[21] In late medieval and early modern Europe, new navigational and calendrical challenges, as well as astrological practices, revived astronomical observation and theory. In the sequel, astronomical observation became increasingly institutionalized, an outstanding example being the observatory granted to Tycho Brahe by Frederick II of Denmark. The facilities allowed the precision of astronomical observations to be substantially improved, owing not only to expensive instrumentation but also to the longer time frames over which systematic observation was possible. The new corpus of observational data laid the foundation for Johannes Kepler's work. He pursued this work within the framework of Copernican heliocentric cosmology, a system which initially did not outperform the Ptolemaic system as regards agreement with observation, but matched it in mathematical elaboration and

---

[20] See Hunger and Pingree (1999).
[21] See Saliba (1994).

facilitated easier calculations.[22] Kepler's work on the orbit of Mars led him to formulate new laws of planetary motion, famously describing planetary orbits by means of ellipses. Kepler's laws constituted the most concise description of the most advanced knowledge of celestial motions on the eve of the invention of the telescope. Later telescopic observations further corroborated the heliocentric hypothesis and Kepler's laws, although stellar parallaxes, which provide the clearest evidence for the earth's annual motion, were not observed before the 1830s.

Large parts of these two strands of knowledge tradition were integrated within a new mathematical-conceptual framework, later referred to as *classical mechanics*. In his *Philosophiae naturalis principia mathematica*, Newton built upon the laws of motion as they were formulated in preclassical terrestrial and celestial mechanics, and thereby incorporated the experiential knowledge they embodied.[23] The new mathematical-conceptual framework was stabilized by its ability to integrate the growing body of experiential knowledge on motion without fundamentally changing its structure. The centerpiece of this new framework was a modified conception of the relation between force and motion. A fundamental structure of elementary mechanical thinking is the motion-implies-force model introduced in the preceding chapter. Roughly, it describes the intuitive expectation that where there is motion, there needs to be a force causing it. Furthermore, elementary experience implies that under usual circumstances, the greater a force the greater its effect. In the Aristotelian-medieval and preclassical conception of motion, this relation between force and motion was expressed in terms of a proportionality of force and velocity: the greater the force, the greater the velocity of the motion it causes. In classical mechanics, by contrast, the model is modified in such a way that only accelerated motions, i.e., motions changing in direction or speed, are in need of a force to explain their occurrence. Uniform motion in a straight line is dynamically equivalent to rest: no forces are present. Force is proportional to acceleration, not to velocity. Thus, in the transition from preclassical to classical mechanics, the deep structure of elementary thinking about motion, the connection between force and motion, is preserved, while its mathematical-conceptual concretization is modified. This modification is the result of a reflective abstraction on the concept of force in the context of an involvement with a body of experiential knowledge hugely expanded as compared to elementary knowledge.

Newton himself did not foresee all conceptual implications of his theory. The conceptual explication of classical mechanics was a process taking centuries and involving the work of many. As we shall see, as late as towards the end of the

---

[22] See, e.g., Johnson (1937, 111–112).

[23] Pierre Duhem is correct, of course, when he points out that Newton could not find his laws by generalizing Kepler's laws, that he could not "extract [them] from experiment" (Duhem 1962, 191) by means of induction. Duhem argues that Newton's set of laws and Kepler's set of laws actually contradict each other and that therefore the one cannot be derived from the other (Duhem 1962, 190–195). It is true that the relation between Newton's mechanics and preclassical mechanics is not one of formal deductivity. The form of reasoning that connects the mutually incompatible but genetically related conceptual systems is non-monotonic and involves content-dependent cognitive structures such as mental models (see below).

nineteenth century, the understanding of the implications of classical mechanics on the notions of space and time changed dramatically. This appears to be a common feature of the development of physical theories that make use of mathematical formalisms. The formalistic structure of the theory is shaped in a process of engagement with the knowledge to be accounted for. The conceptual structure is on the one hand connected to the formalism, but on the other it is rooted in preceding structures. In the development of physical theories over the last centuries, we regularly see conceptual understanding lagging behind the development of formalistic structure. This is clearly an indication that in physics we do not deal with strictly axiomatic theories. Otherwise the concepts would be exclusively fixed by the relations within the theory and not by connotations deriving from structures of thinking that lie outside the theory, structures stemming from older theories or even from entirely different layers of thinking.[24] In accordance with this discrepancy of mathematical structure and conceptual understanding, in Newton's work, the new relation between force and motion becomes more clearly expressed on the mathematical level than on the terminological one. Thus, Newton defines a *vis insita*, an 'inherent force', where in later classical mechanics one would simply speak of *inertia* and avoid the term *force*, since this usage is in contradiction with the new mathematical-conceptual structure that has force proportional to acceleration. In fact, Newton's use of the term 'force' in this instance is reminiscent of the medieval-preclassical concept of impetus and indicates that motion is conceived of as being in need of a moving cause.[25]

We find corresponding reminiscences of earlier conceptions in Newton's concept of space. Therefore, we have to distinguish between Newton's concept and the concept implied by the conceptual-mathematical structure of Newtonian, or classical, physics. The following aspects of Newton's concept of space are described in his unpublished manuscript *De gravitatione...* and in his *Principia*. (They are further discussed by his spokesman Clarke in the Leibniz-Clarke correspondence.)

- *Three-dimensionality and Euclidean structure.* Space is of three dimensions and Euclidean geometry applies throughout.[26] The fact that Euclidean geometry applies is not considered a property of the things filling space, but of space itself, as becomes clear when Newton describes all kinds of geometrical figures to be present in space, albeit insensibly: "In the same way we see no material shapes in clear water, yet there are many in it which merely introducing some colour into

---

[24] A prominent example for the gradual understanding of the conceptual implications of a theory's physico-mathematical formalism, which is itself evolving, is provided by the history of general relativity; see Renn (2007, vols. 1 and 2); see also the next chapter.

[25] The definition of *vis insita* is Definition 3 in the *Principia*, see Newton (1999, 404), in which it is translated as 'inherent force' (another translation is 'innate force', see Newton 1729, 2). Newton's use of the term *force* in this context is discussed by Bernard Cohen in Newton (1999, 96–101).

[26] Three-dimensionality and Euclidicity are not particular to Newton's concept of space, of course; both are accepted properties in ancient, medieval, and early modern theories of space. In the Aristotelian cosmos, for instance, natural motions follow the elements of Euclidean geometrical figures, straight lines and circles. However, while in the Aristotelian case the application of these elements is restricted due to the finiteness of the cosmos, such a restriction does not apply to infinite Newtonian space (see below).

its parts will cause to appear in many ways."[27] The idea that Euclidicity pertains not only to the spatial relations between things in space, but to space itself, is a consequence of Newton's conception of space as something existing independently of anything filling it. This is one of the senses of Newton's space being 'absolute', namely that the relations between objects are "parasitic" upon spatial relations.[28]

- *Infinity.* "Space extends infinitely in all directions."[29] Newton invokes a rationalistic Archytas-type argument to support this thesis: "For we cannot imagine any limit anywhere without at the same time imagining that there is space beyond it."[30] Newton further argues for spatial infinity by referring to the infinity of the Euclidean plane. He describes a triangle whose one side is gradually turned about its common point with the base until it is parallel to the other side, asking: "what was the distance of the last point where the sides met?"[31] Since space is independent from its matter content, Newton considers the possibility of a finite material world within infinite space, in which case mechanics would actually take place in a space of finite extension.[32]

- *Isotropy and homogeneity.* Just like pure Euclidean space, physical space has no distinguished direction and no distinguished places. This directly contradicts Aristotle's conception of a spherically organized cosmos with an 'up' towards the periphery and a 'down' towards the geometrical center. The isotropy and homogeneity of space is a direct consequence of its independence from matter and force.

- *Independence from matter and force.* Space is not structured by its matter content or by the forces acting in it: "[...] in space there is no force of any kind which might impede or assist or in any way change the motions of bodies."[33] This property of space is of crucial importance for the setup of Newtonian mechanics, since it allows for inertial motion, as the continuation of the passage just quoted emphasizes: "And hence projectiles describe straight lines with a uniform motion unless they meet with an impediment from some other source."[34]

---

[27]Newton (1978, 133). This is A.R. and M.B. Hall's translation of Newton's unpublished manuscript *De gravitatione et aequipondio fluidorum*, the Latin original reads: "Ad eundem modum intra aquam claram etsi nullas videmus materiales figuras, tamen insunt plurimae quas aliquis tantum color varijs ejus partibus inditus multimodo faceret apparere" (Newton 1978, 100).

[28]Earman (1989, 11).

[29]Newton (1978, 133).

[30]Newton (1978, 133).

[31]Newton (1978, 134).

[32]Newton (1978, 104, 138). In fact, the assumption of an infinitely extended universe with a homogeneous matter distribution, which appears to be the most natural cosmological extension of Newtonian physics, causes problems for classical theory: the gravitational force on a test-body becomes indeterminate everywhere. Newton himself recognized the problem (and erroneously thought that he could solve it). The problem was rediscovered at the end of the nineteenth century by the astronomer Hugo von Seeliger (1895, 1896). The difficulties for Newtonian theory arising from infinitely extended material universes and the history of approaches to the problem, including Newton's, are discussed in Norton (1999).

[33]Newton (1978, 137).

[34]Newton (1978, 137).

- *Absolute rest.* Space does not move. It is absolutely at rest. This is what justifies to speak of absolute motion when considering motion relative to space. This possibility of defining motion absolutely, i.e., with respect to absolute space, rather than relative to other bodies which may or may not be at rest with respect to absolute space, is a crucial ingredient of Newton's mechanics. It allows him to make sense of the law of inertia in the first place: uniform motion in a straight line is uniform and in a straight line with respect to space at absolute rest. While the law of inertia was part of Descartes' first and second laws of nature,[35] Descartes' definitions of motion render a consistent interpretation of this law impossible.[36]
- *Space becomes detectable when absolute acceleration is considered.* When something is accelerated with respect to absolute space, this can be detected by means of the occurrence of inertial forces. A famous example is Newton's bucket experiment described and interpreted in the Scholium to Definition 8 of his *Principia.*[37] Newton describes the motion of water in a vessel which is quickly spinning about its axis, observing that the water recedes from the center, not when the vessel begins its motion and turns quickly around the water, but later on, when the water participates in the vessel's motion and is therefore at rest with respect to the vessel's walls. The reason for the recession of the water, according to Newton, is its rotational motion with respect to absolute space. He argues that true or absolute motion is to be measured against absolute space, not against the adjacent bodies (as Descartes had claimed), and that the observation of the water being pressed against the vessel's walls provides an instance of the detectability of absolute space.
- *Independence from time.* Space and all of its described features do not change over time.

Newton's arguments (and Clarke's, when arguing in Newton's place) clearly pertain to the philosophical-metaphysical discussions about space described in the previous chapter. There is one innovative feature of the concept of space, however. This is the inertial structure, which served Newton as a proof of the independent existence of space.[38] In the following reception of the concept of absolute space, it was exactly this feature which led to its broad acceptance. It allowed the concept of absolute space to serve as a foundation for a theory of mechanics, which was highly successful in integrating the growing body of knowledge on terrestrial and celestial motions. This emphasis on the mechanical argument is strikingly evident in Leonhard Euler's *Reflections on Space and Time*. Euler makes no attempt to provide a metaphysical

---

[35] Descartes, *Principia*, II, 37–39 (Descartes 1984, 59–61).

[36] Descartes' theory of motion is thoroughly criticized by Newton in his *De gravitatione ...* (Newton 1978, 91–98, 123–131); references to Descartes are less explicit but still clearly discernible in Newton's *Principia*, see below.

[37] Newton (1999, 412–413).

[38] According to Jammer (1954, 114), in Newton's view, the mechanical arguments for the existence of space are subordinate to the theological-metaphysical ones in the sense that their major function is to provide evidence for the relation between absolute space and God. Jammer (1954, 108–114) traces the theological-metaphysical influence on Newton's concept of space back to Jewish cabalistic and Neoplatonic thought, mediated by Henry More and Newton's teacher Isaac Barrow.

argument for the reality of space. Rather, he takes the empirical success of Newtonian mechanics as a proof that reality must pertain to space and time, since they serve as a basis for the concepts of absolute rest and uniform motion in a straight line, which are needed for the foundation of that science. He concludes that any metaphysical derivation that denies this reality can thereby be inferred to be flawed, without further analysis.[39]

> I do not want to enter the discussion of the objections that are made against the reality of space and place; since having demonstrated that this reality can no longer be drawn into doubt, it follows necessarily that all these objections must be poorly founded; even if we were not in a position to respond to them.

Euler argues against the "metaphysicians" by claiming that the reality of absolute space can be demonstrated on purely physical grounds. But a discrepancy remains between what mechanical theory implies and what the concept of absolute space means. While Newtonian mechanics only distinguishes uniform motion in a straight line from accelerated motion, the concept of absolute space provides an absolute standard of rest. This is a consequence of the landmark model. According to this model, which is based on the elementary *dichotomy of movable and unmovable objects*, motion has to be measured with respect to something at rest. Newton and Euler thus preserve, on the conceptual level, a structure of elementary spatial thinking (which also informed the Aristotelian tradition) despite the fact that the laws of mechanics neither require nor support it: these laws offer no means for distinguishing between the states of rest and uniform motion. This discrepancy stems from the mismatch of the modified motion-implies-force model and the landmark model, as we have seen at the end of the foregoing chapter.

This mismatch was only resolved towards the end of the nineteenth century when the concept of *inertial frames* emerged through the work of Carl Neumann, Ludwig Lange, Ernst Mach, and others, who attempted to clarify the implications of classical mechanics for the nature of space. Inspired by Neumann's operational definition of time,[40] Lange provided an operational definition of an inertial system, i.e., a frame of reference within which the laws of classical mechanics hold, by describing the ideal relative motions of three force-free bodies.[41] Lange thus realized an ideal materialization of the reference frames distinguished in classical mechanics, just as astronomers aimed at the best approximative materialization of these frames in terms of astronomical objects. With Neumann and Lange, the Newtonian idea to

---

[39]"Je ne veux pas entrer dans la discussion des objections, qu'on fait contre la réalité de l'espace & du lieu; car ayant démontré, que cette réalité ne peut plus être revoquée en doute, il s'ensuit nécessairement, que toute ces objections doivent être peu solides; quand même nous ne serions pas en état d'y répondre." (Euler 1748, 330, English translation M.S. A German translation is found in Euler 1763, 1–18.) Euler then goes on to state that if one thinks, based on the principle of the identity of indiscernibles, that it is absurd that the different places or parts of space are mutually indistinguishable (as Leibniz did in his correspondence with Clarke), maybe the principle does not hold in general, pertaining to bodies and spirits but not to parts of space.

[40]Neumann (1870). Incidentally, this is the text in which Neumann, in a sort of resublimation of the sublimate landmarks of Newtonian absolute space, introduces the notion of a "Body Alpha."

[41]Lange (1886, 133–141).

define motion with respect to space and time is therefore reversed: space and time are defined with respect to motion.[42]

The result is that the concept of an absolute space defining a standard of rest is eliminated from the foundations of classical mechanics. The landmark model has been adapted to the modified motion-implies-force model: it no longer contains the idea of absolute rest, but only that of rest relative to an inertial system, which is at uniform motion in a straight line relative to an infinity of other inertial systems. A concept of space that does not define a standard of rest or uniform motion but, at the same time, defines a standard of acceleration is in fact something very strange and counter-intuitive, since it does not fit the landmark model. This must be the reason why it only emerged in the later phase of classical mechanics. It was only after classical mechanics had become a firmly established and empirically highly successful scientific theory that a critical analysis of its foundations, the results of which would contradict intuitive models of thinking, became acceptable.

The two contrary positions of Leibniz and Newton can in hindsight be viewed as generalizations of either the relative or the absolute aspect of classical space. Leibniz generalized the relativity with respect to uniform motion to also include accelerated motion and thereby deprived himself of the possibility of relating inertial phenomena to space; Newton generalized the absoluteness of accelerated motion to include uniform motions and thereby claimed the reality of something principally undetectable. The counter-intuitive concept of inertial frames did not constitute a rational choice for Leibniz, Newton, and their contemporaries. Neither did they have the concept of a force field at their disposal to address the question of motion.

Example: The Force Field as a Hybrid Between Objects and Space

Disregarding modern technology, electric and magnetic forces are a fringe phenomenon in everyday life. Although, as we know today, they play a crucial role in the build-up of matter and are ubiquitous in the form of radiation, they usually do not occur as forces between macroscopic bodies. In contrast to the effects of gravitating masses, which add up to cosmic scales, the microscopic electric and magnetic effects are most often neutralized on a macroscopic level. While the falling motion of bodies thus appears 'natural' once a child has become accustomed to it, electric and magnetic forces remain astonishing. Accordingly, the effects of gravitation became central ingredients of natural philosophy (consider Aristotle's anisotropic cosmos, which is fundamentally structured by the directions of the 'natural motions'), while electricity and magnetism remained marginal curiosities for a long time. Experiences with magnetism in the context of the use of the magnetic compass and its practical relevance for seafaring were a major factor in the early modern interest

---

[42]Barbour (1989, 659).

in magnetism for which William Gilbert's work is the most prominent example.[43] But it was only after the invention of the Voltaic Pile, a first type of an electric battery, that a systematic investigation of electric and magnetic forces became possible. Electrical and chemo-electrical experiments now became widespread,[44] a development, which opened a pathway into the investigation of the microcosmos, as Michael Faraday pointed out when he remarked that "[l]ight and electricity are two great and searching investigators of the molecular structure of bodies [...]."[45]

Owing to their marginality in everyday experience, there is no obvious elementary knowledge structure accounting for electric and magnetic forces. The action of forces without material mediation contradicts elementary experience in physical labor. Accordingly, from ancient to modern times, attempts at an explanation often took recourse to invisible substances—'exhalations' or 'effluvia'—that were thought somehow to push or pull the visible bodies.[46] With Newton's success in describing gravitation by means of a universal law and, at the same time, his failure to explain it in physical terms, the counter-intuitive idea of action at a distance became a more reasonable option. The law of gravitation served as a model when Charles Auguste de Coulomb described the electrostatic force between two charged bodies as an action at a distance proportional to the charges (corresponding to the masses in the case of gravitation) and to the inverse square of the distance. The concept of action at a distance, physically unsatisfying but empirically highly successful, provoked different reactions. One possibility was to continue the search for material explanations of forces, which aimed at an elimination of action at a distance. One could, for instance, think of the action of an all-pervading ether, in order to explain electric and magnetic, and, in fact, also gravitational forces.[47] The idea of collisions as the only true interaction between bodies, which is based on the model of the tangible, impenetrable object, is an exemplary, intuitively obvious way to model interaction without recourse to action at a distance. But a further analysis of collision processes reveals problems with this way of reducing forces to matter in motion. If the colliding particles are perfectly rigid, their velocities change instantaneously in the moment of collision and the forces become infinite; if, on the other hand, the particles deform, there must be counter-acting forces within the particles and the reduction of forces to the collision of bodies has failed. Such difficulties may lead to the other extreme view that, on the contrary, there are no perfectly impenetrable bodies, but impenetrability is a derived property and the only fundamental magnitudes are forces. According to this view, repulsive forces explain impenetrability and the concept of force has primacy over that of body. Kant in his *Metaphysical Foundations of Natural Science*

---

[43]On the practical background of Gilbert's work, see Zilsel (2000a).

[44]Chung (1997, 42).

[45]Faraday (1965, Vol. II, 286).

[46]Gilbert gives an account of ancient and later opinions on electricity and magnetism in his *De magnete*, Book II, Chap. 3, before stating his own view (Gilbert 1958, 60–64).

[47]In the case of gravitation, the search for a material explanation continued through post-Newtonian times and well into the twentieth century; see, e.g., van Lunteren (1991).

formulates such a view. A similar view was expressed by Joseph Priestley.[48] A later, less radical conception was that of Laplacian physics, which assumed particles as well as distant forces and attempted to reduce physics to the interactions between the particles of ponderable matter and the imponderable fluids of electricity, magnetism, heat, and light.[49]

Research following the invention of the battery brought about rich experiential knowledge on electrical and magnetic interactions, knowledge that had to be taken into account when speculating about the mediation of these forces.[50] In 1820, Hans Christian Ørsted reported on his discovery of an interaction between a conducting wire and a magnetic field and interpreted his results drawing, among other things, on Kant's theory of matter-constituting forces. Among those who tried to replicate Ørsted's experiment and eventually observed further regularities in electromagnetic experiments was André-Marie Ampère who generalized Ørsted's results on the inter- actions between conducting wires. Ampère explained electromagnetic forces using Augustin Jean Fresnel's concept of a luminiferous ether, the hypothetical carrier of light waves.[51] Among the experimental results obtained by Faraday were the discov- ery of the rotation of a magnetic needle about a conducting wire, electromagnetic induction (i.e., the induction of a current in bodies moving relatively to a magnet), the laws of electrolysis, and the rotation of the plane of polarization of light by mag- netic forces. In the course of changing explanations for the observed effects, Faraday introduced the concepts of 'lines of force' and, in 1845, 'field of force'. Initially, he explained the propagation of forces by means of an 'electro-tonic' state, which he described as a polarization of the molecules of matter. The concrete mechanism of the propagation remained unclear, however, and Faraday seems to have conceived of electrical forces as being manifest in the space around electrified matter particles. Later, Faraday renounced the idea of atomic constituents of matter and described matter as a collocation of forces. In this context, he re-interpreted the lines of force as something having real existence in space and producing effects through its motions and vibrations.[52] Sir William Thomson (Lord Kelvin) further elaborated the theory of the field of force. He eventually arrived at a theory in which atoms were conceived of as stable vortices within a fluid ether and the field of forces was represented by means of motions within that ether.[53]

While the interpretation of the wealth of electromagnetic phenomena in terms of mechanical properties of a fluid ether continued to present fundamental problems, their unified description by means of a mathematical formalism originating in fluid

---

[48]On Faraday's ideas of matter as a 'plenum of powers' being influenced by Priestley's view, and the conflation of Priestley's theory of matter with Roger Joseph Boscovich's, see Harman (1982, 77).

[49]Darrigol (2000, 1), Fox (1974).

[50]For a comprehensive account on the development of electrodynamics in the course of the nine- teenth century, see Darrigol (2000). For a detailed discussion of Ampère's and Faraday's experiments and their relation to theory, see Steinle (2005).

[51]On Ørsted and Ampère, see Harman (1982, 30–32).

[52]Harman (1982, 73–78).

[53]Harman (1982, 82–84).

mechanics advanced. In particular, James Clerk Maxwell succeeded in integrating the current knowledge on electromagnetism in a set of differential equations, which led him to the suggestion that light is an electromagnetic wave. This insight was later corroborated in experiments by Heinrich Hertz, which led to a broad acceptance of the concept of electromagnetic field among continental European physicists. Attempts to interpret the formalism of the field in terms of mechanical motions within an ether continued. Maxwell himself argued that the phenomena of light and heat provided evidence of an ethereal medium filling space and permeating bodies. In particular, he considered the dependence of forces on relative velocities as speaking against an explanation in terms of action at a distance.[54] In fact, even Wilhelm Weber, who described electromagnetic forces by means of a particle-particle-interaction law, extending the electrostatic law to include terms involving relative velocities and accelerations, assumed the transmission of force by an ethereal medium.[55] Maxwell proposed mechanical models such as incompressible fluids moving along tubes formed by lines of force and a cellular ether, in which the rotational direction of vortices is communicated via idle wheels. Yet, he insisted that the mechanical analogy was only imaginary.[56] Various other models involving motions within solid and fluid types of ether were proposed, often accompanied by remarks disclaiming their reality and emphasizing their purely heuristic value. Joseph Larmor for instance, while representing the ether by the model of a rotationally elastic, homogeneous fluid, considered the Lagrangian formalism to be sufficient for stating all that can be physically known.[57] While mechanical modeling of electromagnetic phenomena thus remained a problematical line of work, the description of these phenomena in terms of dynamical equations was highly successful.[58]

No matter what epistemic status was ascribed to the concrete mechanisms of ether dynamics, falling under the mental model of a physical body, the ether had to have a defined state of motion. That is, it either had to be at rest or it had to move in a certain way, possibly being dragged along by ponderable bodies moving through it, such as the planets. Based on the fact that by Maxwell's unification light was an electromagnetic phenomenon and that therefore the electromagnetic and the luminiferous ethers had to be considered one and the same, the conceptualization of the ether's state of motion met with new difficulties as the experiential knowledge concerning relative motions of light and matter expanded.[59] The phenomenon of stellar aberration, the results of the Fizeau experiment on the speed of light in moving water, and the results of Michelson and Morley's experiment on ether drift presented difficulties to a consistent interpretation of the state of motion of the ether. The

---

[54]Maxwell (1890, 527–528).

[55]Harman (1982, 104).

[56]On Maxwell's theories of the field, see Harman (1982, 84–98).

[57]Harman (1982, 101–102).

[58]On the distinction between mechanical models and dynamical systems, see Buchwald (1988, 20–23).

[59]For a detailed account of the history of conceptions of the ether's state of motion in the light of the experiential knowledge of the time, see Janssen and Stachel (2004).

most consequential reaction to this state of affairs was Hendrik Antoon Lorentz's formulation of a theory of electrodynamics that involved charged matter particles on one hand and the electromagnetic field as an independent physical reality on the other. The ether was still conceived as the carrier of the electromagnetic field, but it was deprived of all mechanical qualities—besides being at rest.[60] Lorentz's ether thus introduced an absolute standard of rest into electrodynamics. As was the case for the standard of rest of Newton's absolute space, however, this standard was not detectable. While motion with respect to the ether should have been detectable, for instance by measuring the speed of light, which was constant with respect to the ether, the measured speed of light turned out to be always the same, independent of such motion. The ether thus appeared to act on physical systems in uniform motion relative to it, and to modify the actual dimensions of bodies and the actual duration of processes in exactly such a manner that uniform motion with respect to it was not detectable.

In the theory of special relativity, these so-called Lorentz length contractions and time dilations are no longer interpreted as real modifications of physical objects and processes. They just reflect deviating results of the space and time measurements of observers in different states of motion with respect to the measured system (see the next chapter). The independence of the speed of light from the state of motion of the light source with respect to the observer is postulated as a principle and not, as in Lorentz' theory, a miraculous result of a perfect compensation of different physical effects.

With the theory of special relativity and its postulate of the independence of the speed of light from the state of motion of the light source, the concept of an ether became obsolete. In Einstein's words, the ether lost its last mechanical property—its immobility.[61] What was formerly considered to be effects of the ether now became interpreted as properties of space and time. Thus, Karl Schwarzschild remarked[62]:

> [...] the completely rigid ether stepped out of the circle of the objects that can be influenced and thus can be more closely perceived, so much so that relativity theory became possible, in which the concept of the ether only appears as a spacetime concept deepened by new experience.

Through the prototype of the electromagnetic field, force fields entered physics as a third entity between material bodies and space. The physical field breaks the dichotomy of object and space by displaying characteristics of both. Attributes that make it space-like are that it is infinitely extended (at least in principle), penetrable, and that the idea of motion cannot be applied to it; attributes that make it body-like are that it contains energy (and therefore has mass; inertial mass according to special relativity, inertial and gravitational mass according to general relativity) and that it

---

[60]This is how Einstein put it in retrospect; see Einstein (1922, 11).

[61]Einstein (1922, 11).

[62]"Der völlig starre Äther trat ferner so sehr aus dem Kreis der beeinflußbaren und damit näher erkennbaren Objekte heraus, daß auch die Relativitätstheorie möglich wurde, bei welcher der Begriff des Äthers nur als ein durch neue Erfahrungen vertiefter Raum-Zeitbegriff erscheint." (Schwarzschild 1913, 598, English translation M.S.)

acts on bodies. This kind of entity is not matched by any mental model of elementary cognition. It had to be forced upon human thinking by an accumulation of experiential knowledge which had its sources outside the realm of everyday experience and whose conceptual organization was mediated by an advanced mathematical formalism. Without the corresponding experiential knowledge, this concept of physical field would indeed be a highly irrational thing to conceive of.

The Character of Spatial Knowledge

The knowledge discussed in this chapter is theoretical knowledge. In distinction from the knowledge discussed in the previous chapter, this kind of theoretical knowledge does not simply arise from reflection, but also from an expanding experiential base which challenges the existing theoretical frameworks. The accumulation of experiential knowledge takes place within institutions specifically designed for the purpose of knowledge acquisition[63] such as the Museion of Alexandria, the Uraniborg observatory on the island Hven, and the *Royal Institution* in London, and often by using instruments specifically designed for the purpose of knowledge acquisition such as astronomical instruments and laboratory equipment. The accumulating empirical knowledge is organized in integrative structures based on symbolic and formalistic languages such as numerical coordinates, analytic geometry, calculus, and differential equations. The way the formalistic languages are used is shaped by the experiential knowledge to be integrated. At the same time, the formalistic languages are informed by concepts and have a repercussion on conceptual structures. It is through this interaction of experience, formalism, and concepts that experiential knowledge shapes conceptual structures. In this process of reflection upon the institutionally accumulated empirical knowledge, the mental models, which were based on elementary and practical experience, are transformed. The accumulating knowledge and its formalistic integration thereby bring about and stabilize models and concepts that are highly counter-intuitive. Examples of counter-intuitive knowledge structures encountered in this chapter are:

- *The earth is of a spherical shape.* The idea of a spherical earth violates the *distinction of the vertical direction* in elementary spatial cognition.
- *Uniform motion is purely relative, while accelerated motion is not.* This idea contradicts the elementary landmark model, according to which motion is entirely absolute: there are landmarks that are absolutely at rest. But it also contradicts the theoretical reflection on the landmark model according to which motion is purely relative and only practical considerations define what is considered at rest. The mixture of both is inconceivable!

---

[63] 'Knowledge acquisition' or 'production', depending on whether one wishes to stress the empirical or the constructive aspect of knowledge growth.

- *There are entities that are neither body nor space but combine spatial and bodily attributes.* The concept of physical field without a material carrier contradicts the elementary *dichotomy of object and space.*

The theoretical knowledge resulting from the expansion of experiential spaces has repercussions on different layers of knowledge. Global, geographical coordinates, for instance, attained practical importance in deep-sea navigation. Coastal shipping primarily relies on landmarks. Mediterranean seafaring from the late Middle Ages on could use the magnetic compass complemented by portolan maps displaying compass directions and distances. But for deep-sea navigation knowing one's absolute position is crucial, since in huge regions there are no landmarks and the distances are too large for dead reckoning. After the discovery of electromagnetic radiation, radio navigation (of which satellite navigation is a later development) has become an important tool for spatial orientation at sea.

Theoretical knowledge resulting from the expansion of experiential spaces also has a feedback on theoretical knowledge in general. The insight into the sphericity of the earth, for instance, which became stabilized by the expanding geographical knowledge, had far-reaching consequences for theories of space, as its central role in Aristotelian physics and cosmology illustrates. The success of electrodynamics, to give another example, inspired the electromagnetic worldview according to which all matter should be reducible to fields. Further, the application of the field model to gravitation lay at the foundation of the development of general relativity, as will be discussed in the following chapter.

But theoretical knowledge resulting from the expansion of experiential spaces may also have an impact on meta-theoretical knowledge. This is strikingly demonstrated by the influence of Newton's concept of space on Kant's epistemology. Long before writing the *Critique of Pure Reason*, Kant received the Leibniz-Clarke correspondence and occupied himself with the concept of space, considering aspects of Leibniz's as well as of Newton's conceptions. In the *Critique*, Kant presents space as the pure form of outer intuition and states that[64]

> [w]e can never have a presentation of there being no space, even though we are quite able to think of there being no objects encountered in it.

While space is thus a precondition of experience, rather than being derivable from experience, with matter this is not the case, as Kant explains in his post-critical *Metaphysical Foundations of Natural Science*, in which he strives to provide a sound metaphysical foundation for Newtonian mechanics. In contradistinction to space, matter is an 'empirical concept', i.e., it is in need of perceptually given instances in order to attain objective reality.[65] This epistemic divide between space and matter was not part of Newton's philosophy of space. But the autonomy of Newton's concept of space with respect to the concepts of things in space (matter, force) enabled Kant to put forward such an epistemic separation. Kant clearly argues on the basis of

---

[64] Kant (1996, 78).

[65] On Kant's empirical concept of matter, see Friedman (2001).

a container model of space, even though he does not argue for the reality of this container but only for its necessity in cognition.[66] Kant's epistemic separation of space and matter would not have been possible on the background of Aristotelian physics or general relativity, both representing frameworks in which space is (in very different ways) inseparably intertwined with matter.

References

Abattouy, M., Renn, J., & Weinig, P. (2001). Transmission as transformation: The translation movements in the medieval East and West in a comparative perspective. *Science in Context, 14*, 1–12.

Aristotle, (1987). *Meteorologica*. Aristotle in twenty-three volumes. Cambridge, MA: Loeb Classical Library, Harvard University Press.

Barbour, J. B. (1989). *Absolute or relative motion? Vol. 1: The discovery of dynamics*. Cambridge, MA: Cambridge University Press.

Buchwald, J. Z. (1988). *From Maxwell to microphysics: Aspects of electromagnetic theory in the last quarter of the nineteenth century*. Chicago, IL: University of Chicago Press.

Chung, Y. (1997). Die Entwicklung des Kraftbegriffes im 18. und 19. Jahrhundert und die Entstehung des Feldbegriffes als einer Entität zwischen Raum und Materie. Preprint 68. Berlin: Max Planck Institute for the History of Science.

Clagett, M. (1959). *The science of mechanics in the Middle Ages*. Madison, WI: University of Wisconsin Press.

Cullen, C. (1976). A Chinese Eratosthenes of the flat earth: A study of a fragment of cosmology in Huai Nan Tzu. *Bulletin of the School of Oriental and African Studies, University of London, 39*(1), 106–127.

Damerow, P., Freudenthal, G., MacLaughlin, P., & Renn, J. (2004). *Exploring the Limits of Preclassical Mechanics: A study of conceptual development in early modern science; free fall and compounded motion in the work of Descartes, Galileo, and Beeckman* (2nd ed.). New York: Springer.

Darrigol, O. (2000). *Electrodynamics from Ampère to Einstein*. Oxford: Oxford University Press.

Descartes, R. (1984). *Principles of philosophy*. Dordrecht: Reidel.

Dilke, O. A. W. (1987). The culmination of Greek cartography in Ptolemy. In J. B. Harley & D. Woodward (Eds.), *Cartography in prehistoric, ancient, and medieval Europe and the Mediterranean* (pp. 177–200). Chicago, IL: University of Chicago Press.

Duhem, P. M. M. (1962). *The aim and structure of physical theory*. New York: Atheneum.

Earman, J. (1989). *World enough and space-time: Absolute versus relational theories of space and time*. Cambridge, MA: MIT Press.

Einstein, A. (1922). Ether and the theory of relativity. *Sidelights on relativity* (pp. 3–24). London: Methuen.

Euler, L. (1748). Reflexions sur l'espace et le tems. *Memoires de l'Academie des sciences de Berlin, 4*, 324–333.

---

[66]Compare Kant's statement above to the following contradicting one, made by David Hume in his *Treatise concerning human nature*: "the ideas of space and time are [...] no separate or distinct ideas, but merely those of the manner or order, in which objects exist": "[...] 'tis impossible to conceive either a vacuum and extension without matter, or a time, when there was no succession or change in any real existence" (Hume 2007, 31). Hume here clearly advocates a position-quality concept of space.

Euler, L. (1763). *Vernuenftige Gedanken von dem Raume, dem Orth, der Dauer und der Zeit. Theils aus dem Französischen des Herrn Professor Eulers uebersezt, theils aus verschiednen ungedruckten Briefen dieses berühmten Mannes mitgetheilt; nebst einigen Anmerkungen und einem Versuche einer unparteyischen Geschichte der Streitigkeiten über diese Dinge.* Schwan, Quedlinburg.

Faraday, M. (1965). *Experimental researches in electricity.* New York: Dover.

Fox, R. (1974). The rise and fall of Laplacian physics. *Historical Studies in the Physical Sciences, 4,* 89–136.

Friedman, M. (2001). Matter and motion in the 'Metaphysical Foundations' and the first 'Critique': The empirical concept of matter and the categories. In E. Watkins (Ed.), *Kant and the sciences.* Oxford: Oxford University Press.

Gilbert, W. (1958). *On the magnet.* New York: Basic Books.

Harman, P. M. (1982). *Energy, force, and matter: The conceptual development of nineteenth-century physics.* Cambridge, MA: Cambridge University Press.

Harley, J. B., & Woodward, D. (Eds.). (1987). *The history of cartography, Vol. 1: Cartography in prehistoric, ancient, and medieval Europe and the Mediterranean.* Chicago, IL: University of Chicago Press.

Hume, D. (2007). *A treatise of human nature: A critical edition.* Oxford: Clarendon Press.

Hunger, H., & Pingree, D. (1999). *Astral sciences in Mesopotamia.* Leiden: Brill.

Jammer, M. (1954). *Concepts of space: The history of theories of space in physics.* Cambridge, MA: Harvard University Press.

Janssen, M., Stachel, J. (2004). The optics and electrodynamics of moving bodies. Preprint 265. Berlin: Max Planck Institute for the History of Science.

Johnson, F. R. (1937). *Astronomical thought in Renaissance England: A study of the English scientific writings from 1500 to 1645.* Baltimore, MD: Johns Hopkins University Press.

Kant, I. (1996). *Critique of pure reason: Unified edition; with all variants from the 1781 and 1787 editions.* Indianapolis, IN: Hackett.

Lange, L. (1886). *Die geschichtliche Entwickelung des Bewegungsbegriffes und ihr voraussichtliches Endergebnis: Ein Beitrag zur historischen Kritik der mechanischen Principien.* Leipzig: Engelmann.

Lefèvre, W. (1978). *Naturtheorie und Produktionsweise. Probleme einer materialistischen Wissenschaftsgeschichtsschreibung - Eine Studie zur Genese der neuzeitlichen Naturwissenschaft.* Darmstadt/Neuwied: Hermann Luchterhand.

Mach, E. (1989). *The science of mechanics: A critical and historical account of its development.* Lasalle, IL: Open Court Publ.

Maier, A. (1952). *An der Grenze von Scholastik und Naturwissenschaft: Die Struktur der materiellen Substanz, das Problem der Gravitation, die Mathematik der Formlatituden,* No. 3 in Studien zur Naturphilosophie der Spätscholastik. Rome: Edizioni di Storia e Letteratura.

Malkin, I. (2011). *A small Greek world: Networks in the ancient Mediterranean.* New York: Oxford University Press.

Maxwell, J. C. (1890). A dynamical theory of the electromagnetic field. In W. D. Niven (Ed.), *Scientific papers* (Vol. 1, pp. 526–597). Cambridge, MA: Cambridge University Press.

Neumann, C. (1870). Ueber die Principien der Galilei-Newton'schen Theorie: Akademische Antrittsvorlesung, gehalten in der Aula der Universität Leipzig am 3. November 1869. Teubner, Leipzig.

Newton, I. (1729). *The mathematical principles of natural philosophy.* London: Printed for Benjamin Motte.

Newton, I. (1978). *Unpublished scientific papers of Isaac Newton: A selection from the Portsmouth collection in the University Library.* Cambridge, MA: Cambridge University Press.

Newton, I. (1999). *The 'Principia': Mathematical principles of natural philosophy.* Berkeley, CA: University of California Press.

Norton, J. D. (1999). The cosmological woes of Newtonian gravitation theory. In H. Goenner, J. Renn, J. Ritter, & T. Sauer (Eds.), *The expanding worlds of general relativity* (pp. 271–323). Boston: Birkhäuser.

Omodeo, P. D., & Tupikova, I. (forthcoming). Cosmology and epistemology: A comparison between Aristotle's and Ptolemy's approaches to geocentrism. In M. Schemmel (Ed.), *Spatial thinking and external representation: Towards an historical epistemology of space*. Berlin: Edition Open Access.

Renn, J. (Ed.). (2007). *The genesis of general relativity*. Boston Studies in the Philosophy of Science (Vol. 250). Dordrecht: Springer.

Renn, J., & Schemmel, M. (2006). Mechanics in the Mohist Canon and its European counterparts. In H. U. Vogel, C. Moll-Murata, & G. Xuan (Eds.), *Studies on ancient Chinese scientific and technical texts: Proceedings of the 3rd ISACBRST* (pp. 24–31). Zhengzhou: Daxiang chubanshe.

Saliba, G. (1994). *A history of Arabic astronomy: Planetary theories during the golden age of Islam.* New York: New York University Press.

Schemmel, M. (2006). The English Galileo: Thomas Harriot and the force of shared knowledge in early modern mechanics. *Physics in Perspective, 8*, 360–380.

Schemmel, M. (2008). *The English Galileo: Thomas Harriot's work on motion as an example of preclassical mechanics*. Boston Studies in the Philosophy of Science (Vol. 268). Dordrecht: Springer.

Schemmel, M. (2014). Medieval representations of change and their early modern application. *Foundations of Science, 19*, 11–34.

Schwarzschild. K. (1913). Antrittsrede des Hrn. Schwarzschild. *Sitzungsberichte der Königlich Preussischen Akademie der Wissenschaften*, 596–600.

Seeliger, H. von (1895). Über das Newton'sche Gravitationsgesetz. *Astronomische Nachrichten, 137*, 129–136.

Seeliger, H. von (1896). Über das Newton'sche Gravitationsgesetz. *Sitzungsberichte der Bayerischen Akademie der Wissenschaften, Mathematisch-Physikalsche Klasse, 26*, 373–400.

Steinle, F. (2005). *Explorative Experimente: Ampère, Faraday und die Ursprünge der Elektrodynamik*. Steiner, Stuttgart.

Strabo. (1982–1995). *The geography of Strabo*, reprint edition. Loeb Classical Library. Cambridge, MA: Harvard University Press.

Stückelberger, A., & Graßhoff, G. (Eds.). (2006–2009). *Ptolemaios. Handbuch der Geographie*. Basel: Schwabe.

Szabó, A., & Maula, E. (1982). *Enklima. Untersuchungen zur Frühgeschichte der antiken griechischen Astronomie, Geographie und der Sehnentafeln*. Athen: Akademie Athen.

Talbert, R. (2007). Peutinger's Roman map: The physical landscape framework. In M. Rathmann (Ed.), *Wahrnehmung und Erfassung geographischer Räume in der Antike* (pp. 221–230). Mainz: Philipp von Zabern.

van Lunteren, F. H. (1991). Framing hypotheses: Conceptions of gravity in the 18th and 19th centuries. Ph.D. thesis, Rijksuniversiteit te Utrecht, Utrecht.

Zilsel, E. (2000a). The origins of William Gilbert's scientific method. In D. Raven, W. Krohn, & R. S. Cohen (Eds.), *The social origins of modern science* (pp. 71–95). Dordrecht: Kluwer.

Zilsel, E. (2000b). *The social origins of modern science*. Boston Studies in the Philosophy of Science (Vol. 200). Dordrecht: Kluwer.

Chapter 7
The Decline of an Autonomous Concept of Space

Abstract The renewed revolution of the concept of space in twentieth-century physics can be understood as a process of reflective knowledge integration, an integration of disciplinarily highly structured knowledge. The chapter discusses the loss of autonomy of the concept of space that resulted from the demise of the Newtonian concept. The examples presented are (1) the spacetime of special relativity, which emerged from an integration of knowledge from mechanics and electrodynamics and resulted in a close entanglement of the concepts of space and time; and (2) the spacetime of general relativity, which emerged from the additional integration of knowledge on gravitation and resulted in a close entanglement of the concepts of space and matter (or energy).

Keywords Disciplines · Modern physics · Special relativity · Minkowski spacetime · General relativity

The Object of Study

In the previous chapters we have seen how under more and more specific cultural conditions increasingly general concepts of space emerged. In societies where centralized state administrations took over the social control of space, spatial measures became more standardized and integrated and eventually assumed general arithmetic properties (Chap. 4). In societies where oral and written disputation became a social practice, spatial terms formerly used in specific contexts of action attained abstract meanings defined by their position in more encompassing conceptual systems (Chap. 5). Under specific historical circumstances in early modern Europe, the integration of different historical strands of knowledge culminated in Newtonian mechanics and brought about a concept of space that was not only general but, at the same time, implied the autonomy of space from other physical entities represented by fundamental concepts such as *matter*, *force*, and *time* (Chap. 6). While in its autonomy the space of this conception was similar to the void of ancient atomism, it was clearly not conceived of as *nothing*, but rather as a physical entity in its own

© The Author(s) 2016
M. Schemmel, *Historical Epistemology of Space*, SpringerBriefs in History of Science and Technology, DOI 10.1007/978-3-319-25241-4_7

right, sometimes even as a *substance*, and often as conceptually prior to the things filling space.

The trend for increasingly general spatial concepts under ever more specific cultural conditions did not continue, however, within institutionalized physics and its neighboring disciplines over the course of the twentieth century. It is true that the concepts of space employed in modern physics are more general than the Newtonian concept in that they pertain to theories that are able to integrate a larger corpus of empirical knowledge. An obvious illustration of this fact may be given by referring to general relativity, which contains Newtonian gravitation theory as a limiting case and, in addition, is able not only to predict the advancement of the planets' perihelia as well as the bending of light by gravitation with high precision, but also to describe the spacetime dynamics of massive objects such as galaxy nuclei and, in fact, of the universe in its entirety. Yet, in two important respects the Newtonian concept constitutes the historical acme of the generality of concepts of space: it was thought of as fundamental not only for the theory of mechanics from which it arose, but for the physical world in general, regardless of what was considered to be in that space and what discipline described things in space. It was further considered to be universal in the sense that space was the same everywhere: it was homogeneous and isotropic. This property was closely related to its autonomy from other fundamental concepts; since the distribution of things in space (matter and forces, for example) is obviously not homogeneous, space has to be decoupled from these things in order to be considered homogeneous.

In twentieth century physics, we see these two aspects of generality become inapplicable to the developing concepts of space. While the aspiration to formulate fundamental concepts underlying all of physics has always remained a part of the agenda of theoretical physics, and unification is one of the major challenges of present-day theoretical work, there is no concept of space in twentieth century physics that could consistently be applied to all fields of physics. The same applies to the concepts of time, matter and force. The most advanced concept of space in a well-established theory of modern physics is clearly the one contained in the dynamic spacetime of general relativity, which also plays a central role in modern cosmology. At the same time, this concept of spacetime is not compatible with quantum theory, which up to the present provides us with the most advanced theory of matter and radiation. Thus, quantum field theory usually presupposes a special-relativistic spacetime, and quantum mechanics is mostly done in non-relativistic space. It is unproblematic, of course, to understand fundamental concepts such as *matter* and *space* differently in the different fields of physics. The point is that, if these different usages are understood as resulting from the consideration of limiting cases to a unifying theory,[1] such a unifying theory has not yet been established and we do not know what its concept

---

[1] A limiting case to a theory is understood as the theory that results from the original, more general theory when some dimensional constant of it is taken to be zero, just how special-relativistic spacetime results from general relativity in the limiting case of weak gravitational fields. For a detailed account of limit relations between physical theories, see Ehlers (1986).

of space will look like. There is not even agreement on the way the two fundamental theories of twentieth century physics, quantum theory and general relativity, are to be combined for an advanced understanding of their relation. Is quantizing general relativity the solution? Or can gravitation theory, on the contrary, explain quantum mechanical measurement?[2]

In twentieth century physics the concept of space also lost its generality in terms of autonomy, namely its independence from time, matter, force, and motion, which was a precondition for its universal homogeneity and isotropy. With special relativity space becomes entangled with time such that the way the two may be separated depends on the relative state of motion of the observer and the system under consideration. With general relativity, this spacetime further becomes entangled with matter and force; where the geometry of spacetime is determined by matter (and other forms of energy), and determines the motion of matter and radiation under what was classically considered the gravitational force. This mutual entanglement of spacetime and matter is so close that a separate consideration of matter and spacetime (what is the geometry of spacetime?—how is matter distributed in that spacetime?) can only be done in special cases and only approximatively, while the full theory always demands considering both at the same time. Quantum theory provides further intriguing instances of an entanglement of spatial and material concepts. This may be illustrated by reference to the non-local quantum phenomena often referred to as *quantum entanglement*.[3]

Therefore, when one takes as a topic the decline of an autonomous concept of space, one must study the development of conceptual frameworks underlying the different theories of modern disciplinary physics. Here we shall restrict ourselves to an elementary discussion of the development of special and general relativity. In particular, we shall discuss the following fundamental properties of relativistic space:

---

[2]This latter view has, for instance, been expressed by Roger Penrose (1989, 348–373). A similar view was expressed by Richard Feynman in a letter to Victor Weisskopf dated January 4 to February 11, 1961: "[…] how can we experimentally verify that [gravitational] waves are quantized? Maybe they are not. Maybe gravity is a way that quantum mechanics fails at large distances" (Feynman papers, Box 66, Folder 7, p. 15, Caltech Archives). In current approaches to an integration of gravity with quantum theory, one can still discern the different viewpoints on the nature of spacetime of the different physics communities. Thus, most varieties of string theory (which grew out of quantum field theory) start with a special-relativistic *container-model* spacetime (albeit of ten or more dimensions), within which it is then attempted to unify all fundamental interactions, including gravity, in a quantum theoretical framework. A different approach (closer to the spirit of general relativity) is to 'quantize general relativity', thereby attempting to preserve its *position-quality* view of spacetime (usually referred to as *background independence*). Thus, in Loop Quantum Gravity, a currently successful candidate of this approach, the fundamental objects, the quanta of the gravitational field, are not *in* space. They are nodes in a network of relations (a spin network, technically speaking) and it is quantum superpositions of their aggregates that *constitute* space (Rovelli 2008, 368–369).

[3]On this topic: Blum et al. (forthcoming).

- *Four-dimensionality of spacetime.* The integration of mechanics and electrodynamics under the principle of relativity brought about a new spacetime framework for physics, fusing three-dimensional space with one-dimensional time. This fusion is of such sort that space and time cannot be separated universally but only dependent on the state of motion of a given observer.
- *Dynamicity of spacetime.* The further integration of the gravitational force into the new spacetime framework brought about another fundamental change, rendering spacetime geometry dependent on its content of matter and other forms of energy; spacetime geometry becomes dynamic—"What was the stage in the drama of evolution now joins the troupe of actors."[4]

Example: Four-Dimensionality of Spacetime

The concept of space that was, at the turn from the nineteenth to the twentieth century, broadly assumed to apply universally in the physical sciences—the concept of space from classical physics—was shaped by the study of mechanical systems and their behavior under transformations of space and time coordinates. The concept of inertial frames (see the previous chapter) made it possible to formulate the relativity of uniform motion without having to introduce an absolute standard of rest. The relativity principle could now be formulated by simply stating that the laws of mechanics are the same in all inertial frames. The corresponding coordinate transformations, the so-called Galilean transformations, express the coordinates of one inertial frame in terms of the coordinates of another. Since the two frames may be in relative uniform motion, time enters these transformation formulae, multiplied by relative velocity. But since, in Newtonian physics, time is a universal parameter, and time intervals are the same in all inertial frames, this does not affect spatial measures. The length of a measuring rod, for instance, is independent from the rod's state of motion.

As explained in the previous chapter, Einstein extended the relativity principle to include electrodynamics. Assuming a Lorentzian ether at absolute rest, electrodynamics seemed to violate the relativity principle by distinguishing that inertial frame in which the ether was at rest. This frame was not discernible, however, because Lorentz's and Poincaré's formulae for length contraction and time dilation ensured that electromagnetic phenomena looked the same in all inertial frames. This made it possible to abandon the ether and extend the relativity principle to electrodynamics. The laws of electrodynamics, encapsulated in Maxwell's equations, are the same in all inertial frames, just like the laws of mechanics. But since the laws of electrodynamics contain the velocity of light as a fundamental constant, their relativity implied that the speed of light must be the same in all inertial frames, i.e., it must be independent from the state of (uniform) motion of the light source.

Einstein's theory of special relativity starts exactly from these two principles: the relativity principle ('all laws of physics are the same in all inertial frames'), and the

---

[4] Ashtekar (1988, 2).

principle of the constancy of the speed of light ('the speed of light is independent from the state of motion of the light source').[5] On the face of it, these two principles appear contradictory. If the velocity of light equals c in one inertial system, it must be $c + v$ in an inertial system moving with velocity v with respect to the first one and in opposite direction of the light beam. But this argument presupposes that the Galilean coordinate transformations between inertial frames hold. The two principles at the foundation of special relativity can be reconciled when the transformation rules are modified in an appropriate way.

In order to achieve this, Einstein had to reconsider the rational foundations of the Galilean transformations. In effect, he disentangled two layers of knowledge that were conflated in classical considerations on space and time.[6] On the one hand there is the layer of the operations of measurement. This layer is clearly rooted in practical knowledge and involves concepts such as measuring rods and clocks. In order to apply this knowledge, Einstein had to make basic assumptions concerning the existence of rigid bodies and the possibility of synchronizing clocks, all rooted in elementary mental models and all in accordance with classical physics. On the other hand, there is the layer of theoretical assumptions about the comparison of space and time measures in systems in relative motion, which implies general statements about the structure of space and time. While these assumptions may appear intuitively obvious, Einstein noticed that they are not implied by the assumptions about measurement operations. Giving up the ideas of the independence of length, duration, and simultaneity from the state of motion, Einstein derived new rules for the coordinate transformations between inertial frames. In fact, these rules were already given by the formulae for length contraction and time dilation for motion through the ether. But now they could be re-interpreted as relating results of measurements of lengths and durations in systems in relative motion to each other. The fact that the transformation behavior described by these *Lorentz transformations* was not discovered earlier in the context of mechanics can easily be explained by the small velocities (as compared to the speed of light) usually occurring in mechanics: for small velocities, the Lorentz transformations approach the Galilean transformations.

The Lorentz transformations mix the space and the time coordinates of the systems under consideration. As mentioned above, in Galilean transformations time also occurs, but since it is the universal time which applies in all reference frames, purely spatial and purely temporal intervals are preserved. In Lorentz transformations, by contrast, the space coordinates of one system are generally expressed by a combination of the space and time coordinates of the other. The separation into space and time coordinates becomes dependent on the state of motion and only the measure of a spacetime interval is preserved.

Poincaré and Hermann Minkowski clarified the mathematical consequences of this mixture of space and time, when they developed a four-dimensional formalism to capture special-relativistic kinematics.[7] In the Newtonian case, the length of a path

---

[5]Einstein (1905).

[6]See the discussion in Renn (2007a, 47–48).

[7]See, in particular, Minkowski (1909).

between two points in space depends only on the path. The length of the shortest path between two points, which is the straight line, is called their distance. In Newtonian space, the length of a spatial path is the same in all frames of reference. This means that the three-dimensional, infinitesimal version of (5.1),

$$ds^2 = dx_1{}^2 + dx_2{}^2 + dx_3{}^2, \tag{7.1}$$

is invariant under Galilean transformations. Therefore, the spatial distance between two points is also preserved in Galilean transformations. This is no longer the case for Lorentz transformations. It is the four-dimensional line element

$$ds^2 = c^2 dx_0{}^2 - dx_1{}^2 - dx_2{}^2 - dx_3{}^2, \tag{7.2}$$

that remains invariant under Lorentz transformations; where c is the speed of light and x_0 designates the time coordinate. Therefore, the spacetime 'length' of a spacetime path obtained by integration over the four-dimensional line element (7.2) is preserved. For time-like curves (see below), this is the *proper time*, the time as measured in the rest frame of the moving body. Therefore, we may say that "a good clock is more like a good pedometer than previously thought."[8] The spacetime 'distance' is the *longest* (not the shortest) proper time which corresponds to a straight line in spacetime. (Therefore, in the 'twin paradox', the traveling twin remains younger than the one that stays at home.)

Minkowski introduced the presentation of his spacetime formalism with the famous words[9]:

> Henceforth space by itself, and time by itself, are doomed to fade away into mere shadows, and only a kind of union of the two will preserve an independent reality.

This union can be visualized by a spacetime diagram as illustrated on the left side of Fig. 7.1. It represents the temporal and two spatial dimensions. The paths of all possible light rays passing through a spacetime point (the 'here and now') form a three-dimensional double hypercone, the light cone, which becomes represented as a two-dimensional surface in the diagram. Outside the light cone is the *elsewhere*, which is causally disconnected from the here and now. Events occurring at the here and now cannot influence events at points within the *elsewhere*, and cannot be influenced by them. A curve connecting the here and now to any points in this region of spacetime is called *space-like*. The cone consists of two parts, one directed into the future and one into the past. Within the future cone lie the events (spacetime points) that can be influenced by the here and now, within the past cone lie the events that can influence the here and now. A curve connecting the here and now to any points in the regions within the cone is called *time-like*. Lorentz transformations preserve the causal character of curves:

---

[8] See John Stachel, *Albert Einstein: A Man for the Millenium*[sic]? (http://math.bu.edu/people/levit/AlbertEinstein.pdf, accessed 12 August 2015).

[9] Minkowski (s.a., 75), which is an English translation of the German Minkowski (1909).

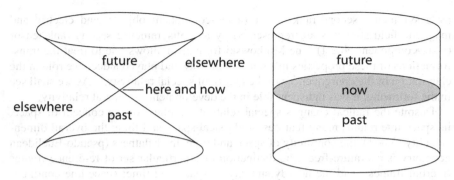

Fig. 7.1 Minkowski spacetime (*left*) and Newtonian spacetime (*right*)

space-like curves remain space-like under Lorentz transformations and time-like curves remain time-like. We thus see that the chrono-geometry of Minkowski spacetime defines its causal structure. Signals at superluminal velocities, if they existed, would violate the causal structure of special-relativistic spacetime.

To understand better what has changed with regards to the Newtonian case, one may draw a similar spacetime diagram representing Newtonian spacetime (Fig. 7.1, right side). In Newtonian spacetime, causal effects may travel instantaneously over arbitrary distances, therefore every event in the past may have an influence on every event now, and every event now may have an influence on every event in the future.[10] Newtonian spacetime therefore naturally separates into space-slices (such as 'now'), which represent three-dimensional spaces of simultaneous events. An inertial frame may now be represented as a congruence of parallel straight lines, each line once and only once cutting through each space-slice of simultaneity. In Minkowski spacetime, inertial frames are equally represented by a congruence of time-like parallels, but the corresponding hyperplanes of simultaneity are different for different frames: the parallels are always pseudo-orthogonal to these planes.[11]

Lorentz transformations can now be understood as (pseudo-)rotations within four-dimensional Minkowski space. This provides an intuitive idea of the mixture of space and time coordinates in Lorentz transformations. This kind of mixture of components pertains, of course, to all physical quantities that are represented by multiple-component objects in Minkowski spacetime. Thus, the energy and momentum of a particle are represented by a four-vector (a four-component object), and the energy, momentum, and internal stresses of an extended body are represented

---

[10]In fact, Newtonian spacetime is obtained from Minkowski spacetime in the limiting case, in which $c \to \infty$. From this, one sees that the metric of Newtonian spacetime is degenerate.

[11]Stachel (1994, 150–151). 'Pseudo-orthogonal', because the metric of Minkowski spacetime is actually a pseudo-metric, which means that it is not positive definite. This can be seen from the mixture of positive and negative terms occurring on the right-hand side of Eq. (7.2). This is why distances may be positive (time-like, according to the convention used in (7.2)), negative (space-like), or zero (light-like). Pseudo-orthogonality of two vectors means that their scalar product computed with the pseudo-metric is zero.

by a symmetric second-rank tensor (a ten-component object), and electric and magnetic field strengths are represented by an antisymmetric second-rank tensor (a six-component object). The Minkowski formalism allows one to treat the transformations of these properties in a convenient way and played a decisive role in the elaboration of the consequences of the theory of special relativity.[12] As we shall see in the following, it was indispensable in the development of general relativity.

Despite the radical changes special relativity implies for the concept of space, its spacetime retains many features of classical space and time: the overall dimensionality $(3 + 1)$, the continuity of space and time, their flatness (pseudo-Euclidean geometry is curvature-free), the distinction of a particular set of reference frames (inertial frames), and the non-dynamicity of space and time: spacetime continues to function as a stage, a set-up background for physics, not a physical object in its own right. Special relativistic spacetime may therefore be understood as based on a modified container model. The modification relates to the integration of space and time. But the separation between the spacetime framework and its matter content (be it particles, fields, or other manifestations of energy) remains intact. This only changes with general relativity.

Example: Dynamicity of Spacetime

For nearly two centuries, Newton's theory of mechanics and gravitation and its elaborations, later referred to as *classical mechanics*, remained successful in integrating ever more precise experiential knowledge about celestial and terrestrial motions. It was only in the middle of the nineteenth century that astronomical observations became precise enough to reveal actual deviations of the celestial motions from the predictions of Newtonian physics: the perihelia of the inner planets advanced with respect to their predicted positions. While an explanation within Newtonian theory was principally possible, it later turned out that these deviations could be explained by a new theory of gravitation which implied radical changes in the relations between the fundamental concepts of space, time, matter and force: general relativity. Yet, it was not the observation of these small deviations from Newtonian theory that caused Einstein and others to reconsider the theory of gravitation. Rather it was internal tensions within the architecture of theoretical physics.

After the advent of the theory of special relativity, it was obvious that Newton's law of gravitation had to be revised in order to fit the new spacetime framework.[13] At the beginning, the revolutionary implications of the adaption of gravitation theory to the new spacetime framework were not obvious and the problem of gravitation could be

---

[12]See, for instance, Laue (1911a).

[13]On alternative theories of gravitation at around the time of the development of general relativity and before, see Renn (2007b, vols. 3 and 4) , in particular the introduction, a revised version of which is Renn and Schemmel (2012).

considered a problem of 'normal science'.[14] In fact, Poincaré and Minkowski both proposed modified laws of gravitation that displayed the transformation behavior demanded by special relativity.[15] Minkowski explicitly stated that his purpose was only to show "that a contradiction to the assumptions of the relativity postulate is not to be expected from the phenomena of gravitation."[16]

The successful example of Maxwell-Lorentzian electrodynamics suggested, however, developing a relativistic field theory of gravitation rather than a particle-particle interaction law, thereby avoiding action at a distance. The example of electrodynamics provided what may be called the *Lorentz model*.[17] This model is a mental structure which describes the physical world as consisting of two kinds of entities: matter and fields. Matter (particles or continuously distributed mass) carries charge, which is a quantity that specifies the way matter interacts with the field. The field is generated by charged matter, which acts as its source. The field is extended in space and acts back on the charged matter. A field equation describes the field as it depends on the charges. It contains a slot for the source of the field (the charges), a slot for the field potential (from which the field can be derived by some differential operator), and a slot for a (second-order) differential operator acting on the potential. Besides the field equation, there is the equation of motion, which determines the motion of charged matter depending on the field strengths.

The analogy between gravity and electrodynamics was not straightforward, however. Gravitation and electromagnetism differ in crucial respects. In particular, while the sources of the electromagnetic field (charges and currents) form a four-vector, the source of the gravitational field is a scalar in Newtonian physics, the gravitational mass. Owing to the way the field equation relates the source to the potential, the potential has to be of the same rank as the source. It was an obvious approach, therefore, to attempt to construct a relativistic field theory of gravitation with a scalar potential. Different options for such a theory were pursued, most notably by Max Abraham and Gunnar Nordström, but both led to theories that ultimately transgressed the special relativistic framework they had started from without providing a new, coherent framework.[18]

Einstein's approach was peculiar in that he combined the quest for a relativistic theory of gravitation with the attempt to extend the relativity principle to include

---

[14]In the sense of Kuhn (1970); see Renn (2007d, 23–24).

[15]Poincare (1906), Minkowski (1908, 98–111); for English translations, see Renn (2007b, vol. 3, 253–285).

[16]Cited after Renn (2007b, vol. 3, 282). Lorentz also discussed a relativistic version of the law of gravitation, noting that it violates the principle of equality of action and reaction; see Lorentz (1910, 1239); for an English translation, see Renn (2007b, vol. 3, 287–301).

[17]Renn and Sauer (2007).

[18]Abraham's approach involved a variable speed of light depending on the gravitational potential and thus broke the mold of Minkowski space. Nordström's final theory connects gravitation to spacetime curvature, so that its *prima facie* Lorentz invariance has a status similar to the Galilean invariance of Lorentzian electrodynamics: the 'real' spacetime measures are not detectible. For accounts on Abraham's and Nordström's work on relativistic field theories of gravitation and Einstein's role in these developments, see Renn (2007c) and Norton (1992) (reprinted as Norton 2007), respectively.

accelerated motion. Now, in accelerated motion inertial forces occur. Einstein's central idea to compensate for this obvious breakdown of relativity was what he referred to as the *principle of equivalence*, which may be formulated thus: "a uniform gravitational field is indistinguishable from a uniform acceleration of a reference frame."[19] Standing in a homogeneous gravitational field is indistinguishable from being uniformly accelerated upwards in force-free space. And freely floating in force-free space is indistinguishable from freely falling in a homogeneous gravitational field.[20] Einstein thus attempted to describe gravitation and inertia as different aspects of one and the same physical entity, an inertio-gravitational field, just as electric and magnetic forces are different components of the electromagnetic field, whose relative strengths depend on the observer's state of motion. Einstein hoped eventually to show that not only all gravitation, but also all inertia originates exclusively from physical bodies and fields and their relative motions, so that, contrary to Newton's conception, in an empty universe there would be no inertia. Einstein later referred to this idea as *Mach's principle*.[21] Mach had criticized Newton's concept of absolute space and had argued for a purely relational theory of motion. According to this conception, inertial forces should be an effect of the relative motions of bodies.[22]

Although several aspects of Einstein's principles later turned out not to be realized in the final version of his theory, the principles served a tremendous heuristic purpose in the theory's development.[23] In particular, since Einstein assumed inertia and weight to be "identical in nature" (*wesensgleich*)[24] and stresses within bodies contributed to their inertia,[25] it was most natural for him to consider the full second-rank stress-energy tensor, rather than a scalar, to be the source of the inertio-gravitational field. The field potential was then also a symmetric second-rank tensor, i.e., a ten-component object $g_{\mu\nu}$, $\mu, \nu = 0, 1, 2, 3$, with $g_{\mu\nu} = g_{\nu\mu}$. This potential Einstein interpreted as the metric tensor, which determines the chrono-geometry of

---

[19]Misner et al. (1973, 189). On different historical formulations of the principle and on the relation between the requirement of uniformity and the restriction to small regions, in particular, see Norton (1989b).

[20]This 'thought observation' hinges on the equality (proportionality, more precisely) of the gravitational and the inertial mass, an equality already assumed in Newtonian physics and experimentally verified with a high precision already in Einstein's times. It is sometimes referred to as the *weak equivalence principle*.

[21]All three principles, the relativity principle, the equivalence principle, and Mach's principle, are briefly discussed in a note by Einstein (1918).

[22]Mach (1989, 271–297). Thus, in view of Newton's bucket experiment (see the previous chapter), Mach remarks: "Try to fix Newton's bucket and rotate the heaven of fixed stars and then prove the absence of centrifugal forces" (Mach 1989, 279). The German has: "Man versuche, das Newtonsche Wassergefäß festzuhalten, den Fixsternhimmel dagegen zu rotieren und das Fehlen der Fliehkräfte nun nachzuweisen" (Mach 1988, 252).

[23]'General relativity' is actually a misnomer, as we shall see below. On the discrepancy between the heuristic ideas that led Einstein to pursue general relativity and the actual implications of the final theory, see, in particular, Janssen (2014). On Mach's principle and general relativity, see Barbour and Pfister (1995).

[24]Einstein (1918, 241).

[25]As Max Laue's work had shown; see, in particular, Laue (1911b).

spacetime. Thus, in extension of the Gaussian line element (5.2),

$$ds^2 = \sum_{\mu,\nu=0}^{3} g_{\mu\nu} dx_\mu dx_\nu, \qquad (7.3)$$

where the $g_{\mu\nu}$ are functions of the coordinates x_μ, $\mu = 0, 1, 2, 3$. Just as Gaussian curved surfaces look Euclidean locally, the curved spacetime of general relativity locally looks Minkowskian, i.e., like special-relativistic spacetime.

There are various ways in which Einstein may have arrived at the insight that the gravitational potential at the same time determines the curved geometry of space-time.[26] For realizing this radical reconceptualization of gravitation, Einstein could again draw on the resources of classical physics: the representation of forced motions in terms of constraints, which are taken care of in the context of a variational principle from which the equations of motion are derived (associated with the names of Jean-Baptiste d'Alembert and Joseph Louis Lagrange, among others); and the theory of geodesics (extremal lines) in curved spaces in current mathematics.[27] The description of gravitation in terms of spacetime curvature had a radical consequence for the conceptualization of force in Einstein's nascent theory. The effects of gravitation had so far been described, in Newtonian physics as well as in the field theoretic formulations of gravitation, on the basis of the modified motion-implies-force model: a gravitational force pulled matter, accelerated it and thereby made it deviate from its inertial path through spacetime. Now, gravitation sided with inertia, not with force: the natural, free-falling motion of matter was geodesic motion within the inertio-gravitational field. Gravitation ceased to be a force; it became an aspect of the structure of spacetime that determines force-free motion.

It remained for Einstein to formulate the field equations of the inertio-gravitational field, a task that would take him three more years.[28] The difficulty consisted in finding equations that came as close to Einstein's principle-borne aims as possible (covariance of the field equations under coordinate transformations as general as possible) while at the same time obeying more conservative demands of physical consistency (Newtonian limit, energy and momentum conservation). The reorganization of the knowledge resources of classical physics and special relativity led to results that contradicted several of Einstein's expectations. In particular, the Lorentz model, which lay at the basis of Einstein's search, was modified in its application to the inertio-gravitational field.

The Lorentz model splits interactions into cause and effect in a particular manner: charges act as sources conditioning the field via the field equation, and the field acts back on charges via the equation of motion. When there are other forces acting on

---

[26] For different conjectures on how Einstein realized in 1912 that he needed to represent gravitation in terms of spacetime curvature, see Stachel (1989b) and Blum et al. (2012).

[27] Renn (2007a, 69).

[28] Pioneering articles on this subject are Stachel (1989a) and Norton (1989a). A comprehensive, state-of-the-art account on this episode in the history of science, based on a meticulous analysis of Einstein's working notes, is given by the various contributions to Renn (2007b, vols.1 and 2).

matter, their dynamic equations can be dealt with separately. In general relativity this splitting no longer applies.[29] First, since the inertio-gravitational field describes the geometry of spacetime, it figures in the description of the source: the distribution of matter (and other forms of energy) in space and time. In particular, it figures in all equations governing the dynamics of matter, not only those related to the inertio-gravitational field. But the spacetime geometry is only given as a solution to the field equation. As a consequence, gravitational and non-gravitational dynamics become entangled.[30] Second, the equation of motion is no longer independent from the field equation and can, under certain conditions, even be derived from it. This deviation from the Lorentz model reflects a fundamental change in the relation between field and source. In fact, the gravitational field acts as its own source, a circumstance that is mathematically reflected in the non-linearity of the field equation.

Both modifications of the Lorentz model in the context of general relativity amount to a close entanglement of spacetime geometry on one hand and matter and other forms of energy on the other. They remain separate concepts, but in the description of their dynamics they are most closely entangled, more closely even than field and source in electrodynamics. Moritz Schlick thus adapted Minkowski's above-quoted statement on the union of space and time in special relativity to general relativity by including matter into the union: "the predicate of reality only befits an indissoluble union of space, time, and matter."[31]

Let us consider how general relativistic spacetime compares to the Newtonian and special relativistic spacetimes illustrated in Fig. 7.1. Matter and energy condition the chrono-geometry of spacetime via the field equation. Locally spacetime looks Minkowskian, but otherwise the causal structure of spacetime is not fixed in advance. Instead of spacetime presenting a stage for physical processes, as is the case in Newtonian and special relativistic physics, general relativistic spacetime takes part in the physical processes, and it does so in a manner that makes it impossible to treat the dynamics of spacetime separately from the other dynamical processes. The spacetime chrono-geometry has a feedback on matter and energy. It determines the inertio-gravitational field, which tells matter how to move. Point particles naturally move along geodesics in the inertio-gravitational field, as long as no force (other than gravity—gravity no longer counts as a force) makes them deviate from their geodesic. In Newtonian and Minkowski spacetimes, geodesic motion was represented by straight lines, the congruences of lines representing inertial frames. Now, corresponding to the absence of a fixed chrono-geometry, there is no longer a fixed inertial structure. Gravitation has made the inertial structure dynamic. Inertial frames no longer play a central role, instead, motion along geodesics is singled out.[32]

---

[29]Renn and Sauer (2007, 293).

[30]Stachel (1994, 159–160).

[31]"[…] nur einer unauflöslichen Einheit von Raum Zeit und Stoff noch das Prädikat der Wirklichkeit zukomme" (Schlick 1921, 103, English translation M.S.). The original remark is in Schlick's book on *Raum und Zeit in der gegenwärtigen Physik* (Schlick 2006, 255–256).

[32]The straightest lines in spacetime are defined by the affine structure, which is, in general relativity, determined by the chrono-geometry.

The difference is no longer between uniform and accelerated motion, but between force-free motion along the geodesic and forced motion along a non-geodesic path through spacetime. The planets are not constantly pulled away from their inertial straight paths through space by the Sun, they are following their straightest possible path in a spacetime warped by the mass of the sun.

In Newtonian and in special-relativistic spacetime the modified landmark model controls the determination of motion. 'Modified', because motion can be determined only *modulo* uniform motion. Coordinates serve as sublimate landmarks. In general relativity, by contrast, motion is determined with respect to something physically very real: the metric field. It makes therefore good sense to regard this as a re-introduction of an ether.[33] This ether is not of the old, mechanical type. The ether itself is a field. It is the field that determines the inertial motions of matter. And it is dynamical, because it is, at the same time, the gravitational field. This then also explains how space can act on matter. In Newtonian physics, and also in special relativity, the fixed inertial spacetime structure may be said to act on accelerated bodies—the manifestation of this action are the inertial forces[34]—but there is no reaction upon it. In general relativity the relation becomes symmetrical. Spacetime has become a physical field that acts on bodies and is acted upon by them. The loss of a fixed background structure in general relativity also implies that it no longer makes sense to speak of relative motion between bodies. Motion is always local, with respect to the metric field. In the case of symmetries within the metric field, it is coordinates again that play the role of sublimate landmarks. In the case of spacetimes without symmetries, by contrast, it is possible to define intrinsic coordinates, which are physical landmarks provided by the inertio-gravitational field itself.[35]

The Character of Spatial Knowledge

The spatial knowledge discussed in this chapter is a particular kind of theoretical knowledge, knowledge that develops only in disciplinarily structured science.[36] The knowledge is characterized by a hierarchy of divisions into areas that display specific knowledge structures comprising area-specific concepts, models, and methods. At the same time, different areas are connected by the overlap of certain concepts, models, and methods. In particular, fundamental concepts such as space, time, energy, matter, and force relate the different areas, without necessarily being understood in the same way. Areas may further be connected by objects of study whose treatment requires

---

[33]Einstein (1922), Weyl (1924).

[34]Think of Newton taking the outcome of the bucket experiment as evidence for the existence of absolute space.

[35]Bergmann and Komar (1972).

[36]On the differentiation of scientific disciplines from the late eighteenth to the early twentieth centuries, for the case of the physical sciences in Germany, see Stichweh (1984) and Jungnickel and MacCormmach (1986).

specific knowledge from more than one area. The knowledge structures within these areas are comparatively stable over periods of time, but knowledge integration across area-boundaries leads to fundamental changes of structure.

The boundaries between the areas variously shift over the course of time, resulting in knowledge integration and disintegration, but overarching theories remain a challenge. Thus, the theory of special relativity resulted from the integration of mechanics and electrodynamics into a unified spacetime framework. This led to the temporary disintegration of gravitation, which had formerly been a part of mechanics. The re-integration of gravitation, mechanics and electrodynamics in a unified spacetime framework brought about general relativity. Quantum mechanics, which had emerged from the consideration of problems on the boundary between thermodynamics and electrodynamics, and further integrated knowledge from mechanics, faced the challenge of integrating relativistic field theory. The integration of special relativistic electrodynamics into a quantum framework—quantum electrodynamics—left the gravitational force—general relativity—standing alone again.[37] In this sub-disciplinary landscape, the two theories of relativity play quite different roles. Special relativity provides the spacetime framework for a large number of sub-disciplinary fields, while general relativity, albeit the more fundamental theory, is comparatively isolated.

The disciplinary organized knowledge is represented by means of highly specialized technical languages, often employing symbol systems, in particular mathematical formalisms. Empirical knowledge is systematically produced in various subfields. The knowledge resources in their disciplinary configuration define a space of possible transformations and thereby condition the outcome. This means that even in cases in which, historically, the development relies on the particular contribution of a single individual, the configuration of knowledge conditions the outcome of the transformation and thereby imposes certain conceptual changes on the protagonists. The invention of general relativity is no exception to this. At first sight it may appear contingent upon Einstein's peculiar insistence on the incorporation of the equivalence principle in a relativistic theory of gravitation, and his lonely work ensuing from it. Nevertheless, granting the necessity of consolidating gravitation with relativity, and given the knowledge resources of classical mechanics, one is almost inevitably led to spacetime curvature. Thus, despite Nordström's more conservative approach to a relativistic theory of gravitation, his final theory exhibits a curved spacetime, as was later demonstrated.[38] One can conduct very basic arguments, demanding energy conservation, deriving the equivalence principle from it, and then showing that in

---

[37] The observation of this latter shift of frontier—from a divide between quantum mechanics and field theory to one between quantum field theory and general relativity—is a result of research done by Alexander Blum, see Blum and Rickles (forthcoming). The very synoptic outline given in this paragraph neglects, among other things, the nuclear forces that also played an important role in the history of twentieth-century physics.

[38] Einstein and Fokker (1914). Historically, this result was again a consequence of Einstein's intervention; see Norton (1992). Abraham's introduction of a variable line element is another point in case; see Renn (2007c, 311–312).

special relativity this inevitably leads to curved spacetime geometry.[39] Theories with a tensor potential starting off in a flat Minkowski spacetime, too, turn out to exhibit a curved spacetime, once their inconsistencies are removed.[40] One may thus conceive of very different historical pathways, probably distributing innovative contributions among more individuals, and combining the classical resources in different temporal order, all eventually leading to a theory very similar to general relativity[41]—or maybe, much less probably, directly to a theory of quantum gravity?

References

Ashtekar, A. (1988). *New perspectives in canonical gravity*. Napoli: Bibliopolis.

Barbour, J. B., & Pfister, H. (Eds.). (1995). *Mach's principle: From Newton's bucket to quantum gravity*. Einstein studies (Vol. 6). Boston: Birkhäuser.

Bergmann, P., & Komar, A. (1972). The coordinate group symmetries of general relativity. *International Journal of Theoretical Physics, 5*, 15–28.

Blum, A., & Rickles, D. (Eds.). (forthcoming). *Quantum gravity in the first half of the twentieth century: A source book*. Berlin: Edition Open Access.

Blum, A., Renn, J., & Schemmel, M. (forthcoming). Experience and representation in modern physics: The reshaping of space. In M. Schemmel (Ed.), *Spatial thinking and external representation: Towards an historical epistemology of space*. Berlin: Edition Open Access.

Blum, A. S., Renn, J., Salisbury, D. C., Schemmel, M., & Sundermeyer, K. (2012). 1912: A turning point on Einstein's way to general relativity. *Annalen der Physik, 524*(1), A11–A13.

Ehlers, J. (1986). On limit relations between, and approximate explanations of, physical theories. In R. Barcan Marcus, G. J. Dorn, & P. Weingartner (Eds.), *Logic, methodology and philosophy of science VII* (pp. 387–403). Amsterdam: North-Holland.

Einstein, A. (1905). Zur Elektrodynamik bewegter Körper. *Annalen der Physik und Chemie, 17*, 891–921.

Einstein, A. (1918). Prinzipielles zur allgemeinen Relativitätstheorie. *Annalen der Physik, 55*, 241–244.

Einstein, A. (1922). Ether and the theory of relativity. *Sidelights on relativity*. London: Methuen.

Einstein, A., & Fokker, A. D. (1914). Die nordströmsche Gravitationstheorie vom Standpunkt des absoluten Differentialkalküls. *Annalen der Physik, 44*, 321–328.

Janssen, M. (2014). 'No success like failure': Einstein's quest for general relativity, 1907–1920. In M. Janssen & C. Lehner (Eds.), *The Cambridge companion to Einstein* (pp. 167–227). Cambridge, MA: Cambridge University Press.

Jungnickel, C., & MacCormmach, R. (1986). *Intellectual mastery of nature: Theoretical physics from Ohm to Einstein*. (2 vols.). Chicago, IL: University of Chicago Press.

Kuhn, T. S. (1970). *The structure of scientific revolutions* (2nd ed.). Chicago, IL: University of Chicago Press.

Laue, M. (1911a). *Das Relativitätsprinzip*. Braunschweig: Vieweg.

Laue, M. (1911b). Zur Dynamik der Relativitätstheorie. *Annalen der Physik, 35*, 524–542.

---

[39] Misner et al. (1973, 177–191).

[40] Misner et al. (1973, 424–425).

[41] One such counter-factual scenario assumes the implementation of the equivalence principle in Newtonian science, leading to a form of classical mechanics that involves an inertio-gravitational field curved in spacetime, so that the step to general relativity becomes almost trivial, once special relativity arrives (Stachel 2007). See also Renn and Stachel (2007), who discuss the convergence of Hilbert's work on the *foundations of physics* with Einstein's theory.

Lorentz, H. A. (1910). Alte und neue Fragen der Physik. *Physikalische Zeitschrift, 11,* 1234–1257.

Mach, E. (1988). *Die Mechanik in ihrer Entwicklung: Historisch-kritisch dargestellt.* Berlin: Akademie Verlag.

Mach, E. (1989). *The science of mechanics: A critical and historical account of its development.* Lasalle, IL: Open Court Publ.

Minkowski, H. (1908) Die Grundgleichungen für die elektromagnetischen Vorgänge in bewegten Körpern. *Nachrichten der Königlichen Gesellschaft der Wissenschaften zu Göttingen,* 53–111.

Minkowski, H. (1909). Raum und Zeit. *Physikalische Zeitschrift, 10*(3), 104–111.

Minkowski, H. (s.a.). Space and time. *The principle of relativity: A collection of original memoirs on the special and general theory of relativity* (pp. 75–91). New York: Dover.

Misner, C. W., Thorne, K. S., & Wheeler, J. A. (1973). *Gravitation.* New York: Freeman.

Norton, J. D. (1989a). How Einstein found his field equations, 1912–1915. In D. Howard & J. Stachel (Eds.), *Einstein and the history of general relativity* (pp. 101–159). Boston: Birkhäuser.

Norton, J. D. (1989b). What was Einstein's principle of equivalence? In D. Howard & J. Stachel (Eds.), *Einstein and the history of general relativity* (pp. 5–47). Boston: Birkhäuser.

Norton, J. D. (1992). Einstein, Nordström and the early demise of scalar, Lorentz covariant theories of gravitation. *Archive for History of Exact Sciences, 45,* 17–94.

Norton, J. D. (2007). Einstein, Nordström and the early demise of scalar, Lorentz covariant theories of gravitation. In J. Renn (Ed.), *The genesis of general relativity* (Vol. 3, pp. 413–487). Dordrecht: Springer.

Penrose, R. (1989). *The emperor's new mind: Concerning computers, minds, and the laws of physics.* New York: Oxford University Press.

Poincaré, H. (1906). Sur la dynamique de l'électron. *Rendiconti del Circolo Matematico di Palermo, 21,* 129–175.

Renn, J. (2007a). Classical physics in disarray: The emergence of the riddle of gravitation. In J. Renn (Ed.), *The genesis of general relativity* (Vol. 1, pp. 21–80). Dordrecht: Springer.

Renn, J. (Ed.). (2007b). *The genesis of general relativity.* Boston Studies in the Philosophy of Science (Vol. 250). Dordrecht: Springer.

Renn, J. (2007c). The summit almost scaled: Max Abraham as a pioneer of a relativistic theory of gravitation. In J. Renn (Ed.), *The genesis of general relativity* (Vol. 3, pp. 305–330). Dordrecht: Springer.

Renn, J. (2007d). The third way to general relativity: Einstein and Mach in context. In J. Renn (Ed.), *The genesis of general relativity* (Vol. 3, pp. 21–75). Dordrecht: Springer.

Renn, J., & Sauer, T. (2007). Pathways out of classical physics. In J. Renn (Ed.), *The genesis of general relativity* (Vol. 1, pp. 113–312). Dordrecht: Springer.

Renn, J., & Schemmel, M. (2012). Theories of gravitation in the twilight of classical physics. In C. Lehner, J. Renn, & M. Schemmel (Eds.), *Einstein and the changing worldviews of physics* (pp. 3–22). Boston: Birkhäuser.

Renn, J., & Stachel, J. (2007). Hilbert's foundation of physics: From a theory of everything to a constituent of general relativity. In J. Renn (Ed.), *The genesis of general relativity* (Vol. 4, pp. 857–973). Dordrecht: Springer.

Rovelli, C. (2008). *Quantum gravity.* Cambridge, MA: Cambridge University Press.

Schlick, M. (1921). Kritizistische oder empiristische Deutung der neuen Physik? *Kant-Studien, 26,* 96–111.

Schlick, M. (2006). *Über die Reflexion des Lichtes in einer inhomogenen Schicht. Raum und Zeit in der gegenwärtigen Physik,* Gesamtausgabe (Abt. 1, Bd. 2). Wien: Springer.

Stachel, J. (1989a). Einstein's search for general covariance, 1912–1915. In D. Howard & J. Stachel (Eds.), *Einstein and the history of general relativity* (pp. 63–100). Boston: Birkhäuser.

Stachel, J. (1989b). The rigidly rotating disk as the 'missing link' in the history of general relativity. In D. Howard & J. Stachel (Eds.), *Einstein and the history of general relativity* (pp. 48–62). Boston: Birkhäuser.

Stachel, J. (1994). Changes in the concepts of space and time brought about by relativity. In C. C. Gould & R. S. Cohen (Eds.), *Artifacts, representations and social practice: Essays for Marx Wartofsky* (pp. 141–162). Dordrecht: Kluwer.

Stachel, J. (2007). The story of Newstein or: Is gravity just another pretty force? In J. Renn (Ed.), *The genesis of general relativity* (Vol. 4, pp. 1041–1078). Dordrecht: Springer.

Stichweh, R. (1984). *Zur Entstehung des modernen Systems wissenschaftlicher Disziplinen: Physik in Deutschland 1740–1890*. Frankfurt: Suhrkamp.

Weyl, H. (1924). Massenträgheit und Kosmos. Ein Dialog. *Die Naturwissenschaften, 12*(11), 197–204.

Chapter 8
Concluding Remarks

Abstract The chapter summarizes some key insights of this study by highlighting developmental strands that connect different forms of spatial knowledge. It closes by positioning the presented approach within the larger field of knowledge studies, in particular arguing for the potential of this approach to consolidate two aspects which are widely perceived to work in opposite directions: the insight that knowledge depends on culture and history on one hand, and the aspiration for a rational foundation of knowledge on the other.

Keywords Forms of knowledge · Experience · Reflection · History of science · Rationality

This book began by raising questions about the epistemic status of spatial cognition. What is the relation between predetermined cognitive structures and experience? To what extent are the structures of spatial cognition universal and how do they depend on cultural conditions? The argument underlying the book was that it is only by studying the history of spatial thinking that the epistemic status of spatial knowledge can be assessed. An attempt was then made to substantiate this claim by discussing different aspects of the historical development of spatial knowledge and analyzing the epistemic status of the related structures of spatial thinking. In particular, we encountered the following forms of space:

- *Naturally conditioned space* is structured by elementary mental models controlling action and perception such as the permanent object model and the landmark model (Chap. 2).
- *Culturally shared space* is represented in language, culturally conditioned actions and cultural artifacts, and builds upon the mental structures of naturally conditioned space, endowing them with cultural meaning (Chap. 3).
- *Administratively controlled space* is represented by measuring tools, arithmetic and linguistic symbols, and schematic drawings, and adds metric significance to structures of the previous forms of space (Chap. 4).
- *Mathematically reflected space* generalizes metric structures through abstraction, using diagrams, formalized language, and other symbol systems for its representation (Chap. 5).

© The Author(s) 2016   107
M. Schemmel, *Historical Epistemology of Space*, SpringerBriefs
in History of Science and Technology, DOI 10.1007/978-3-319-25241-4_8

- *Philosophically reflected space* generalizes linguistically represented elementary structures by elevating them to the rank of principle and exploring the consequences (Chap. 5).
- *Empirically and disciplinarily imposed space* results from the integration of knowledge acquired by systematic observation and experimentation employing conceptual-mathematical formalisms (Chaps. 6 and 7).

A central concern of this book was to show in which ways these forms of space are related from their origin. As we have seen, the occurrence of each new form of space depends on specific socio-cultural conditions. Its concrete realization, by contrast, does not solely depend on these conditions, but also on existing cognitive structures and the addition of further experience. This does not mean that the specific expressions of the forms of spatial knowledge are all connected through developmental relations. Each form may find different expressions in different societies and at different historical times, so that the graph of developmental relations is widely ramified. It also does not mean that the different forms of space only represent historically succeeding stages. They also represent forms of thinking that are simultaneously present within single societies. Different forms of spatial knowledge are shared either by the entire society, or by specialized groups, and may affect each other. Within different societies, they co-exist in varied manifestations, each society displaying its unique spectrum of expressions of spatial thinking.

Nevertheless, within this diversity of cultural expressions of spatial thinking and the complex network of their developmental relations, one can identify cognitive structures that run through the different forms of spatial knowledge like themes. Here we have described these structures in terms of mental models and identified, in particular, the permanent object model and the landmark model. We have further identified structural aspects of these models, such as the *dichotomy of objects and spaces* or the *dichotomy of movable and unmovable objects*, and traced their transformations through the different forms of spatial knowledge. Let us here summarize some results of the previous chapters by giving a condensed, synoptic review of three such transformative strands. After that we shall elaborate on some epistemological implications of these results.

Consider, as a first example, the *dichotomy of objects and spaces*. In its origin, this cognitive structure is part of the sensorimotor intelligence of animals and humans. In humans it develops in the early phase of ontogenesis through close interaction of the infant with the environment. The structure thus reflects experiences made in this interaction. But once the schema of the permanent object is formed, the structure no longer relies on further experience. It has become a premise for further experience. At the same time, its application depends on the specific context of action, and there may be experiences that modify the dichotomy, for instance when fluid bodies are experienced.

As a part of sensorimotor intelligence, the dichotomy of objects and spaces is not reflected upon. Nevertheless, in human language the dichotomy is reflected in various ways. This is the basis upon which, in specific cultural circumstances, the dichotomy may become the object of systematic reflection. In this context it may even be elevated to the level of general principle, as happens in ancient Greek atomism. This is when general concepts form, such as that of a void or of matter. In the context of such reflection, which detaches the structure from any concrete context of action, fundamental questions arise, for instance, whether the void is infinitely extended.

In case a society provides the appropriate institutional context, such reflection about space may be further promoted and become part of a complex tradition of theoretical thinking. In particular, it may be brought into contact with experiential knowledge accumulated in institutions, which goes far beyond the elementary experiences the dichotomy of objects and spaces is built upon. When this experiential knowledge is processed by means of mathematical formalisms, as increasingly was the case in Europe beginning in early modern times, knowledge structures may form that fundamentally contradict the original dichotomy, as is the case with the concept of a force field, a kind of hybrid between objects and space.

Consider the *dichotomy of movable and unmovable objects* as a second example. This cognitive structure is also, in its origins, part of the sensorimotor intelligence of animals and humans. It reflects experience when navigating through an environment and, in turn, constitutes the basis for successful orientation by means of landmarks. Again, from this case we see how the closure of the group of experiences made in the context of a particular type of action (e.g. movement through the environment) transforms experiential structures into premises for further experience.

While the cognitive structure remains unconscious in sensorimotor behavior, cultural systems of navigation and orientation imply a first step of reflection. These systems use spoken language and other material means to represent spatial experience so that it can be communicated and shared, thus accumulating over generations. The cultural systems function by building on the shared structures of elementary spatial cognition. At the same time, they may modify such structures, as the reversal of the roles of landmarks and the self with respect to motion and rest in the case of Carolinean navigation shows. The same case also shows how learned practices may have a repercussion on intuition and modify sensorimotor behavior.

The landmark model is further modified when institutionally accumulated knowledge is integrated by a process of reflection. Thus, when the known geographical space expands and knowledge of locations is combined with local astronomical knowledge, a global system of coordinates may be established within which actual landmarks can be related to numerically defined points. On cosmological scales, new knowledge about the motion of celestial bodies, including the earth, leads to fundamental questions about motion and rest that do not occur in practical contexts. In classical mechanics, which resulted from a reflective integration of a large part of the knowledge on mechanics and astronomy that had accumulated up to early modern times, any system of landmarks can be replaced by any other system in uniform motion relative to it, a very counter-intuitive modification of the landmark model. In the context of general relativity, by contrast, the question of relative or absolute motion is recast on the basis of the field concept. Motion is no longer to be conceived of as relative to distant bodies, but as relative to the local inertio-gravitational field.

Consider the *dependency of the effort on the path taken* as a third and last example. Approaching a goal head-on following a straight line is part of sensorimotor intelligence. It does not presuppose any reflection on what is the shortest path. In fact, when there are obstacles along the straight path, or when another path to the same goal can be taken with less effort, the other path may be chosen. This elementary knowledge structure is preserved in many cultural practices of navigation. When sailing, for instance, riffs, winds and currents may make deviations from the straight path preferable or necessary (as with tacking), but disregarding such circumstances the straight path remains the natural choice, as Micronesian navigational practice exemplifies. Again, the practice is not derived from an abstract notion of the shortest line between two points. In fact, concepts of distance in practical contexts may rely on the time it takes to get from one place to another or on other concrete aspects of the practical procedures, and an independent identification of the geometrically shortest path is usually completely useless in these contexts. Different practices in different terrains and on different spatial scales may therefore lead to different practical concepts of length or distance.

The different concepts of length and distance can only become unified when their external representations are brought into relation. This happens when comprehensive systems of spatial units develop in the administrative institutions of early civilizations. Elementary structures of spatial cognition are then reproduced on the level of (proto-)arithmetical and (proto-)geometrical representations, leading to their metrization, for instance when the additivity of lengths, areas, and volumes becomes a fundamental principle in the handling of spatial problems. Similarly, the use of measuring tools materially reflects the mental structure of the conservation of length, area, and volume with time, place and position.

Theoretical reflections on spatial practices such as measuring and drawing bring about a further generalization of concepts such as distance, length, area, and volume. These concepts may, for instance, be implicitly defined by stating the relations of their sizes after operations of addition or subtraction, as happens with the Euclidean concept of magnitude in the common notions in Book I of the *Elements*. In the tradition of deductively presented geometry, the fact that there always exists one and only one shortest line connecting any two points, the straight line, can even be proven from basic assumptions. The second-order representation of geometrical knowledge by means of a technical language, including meta-linguistic terms like 'postulate', 'theorem', and 'proof', (as contrasted with first-order representations such as drawn figures) implies the potential for further reflective abstraction. This potential was realized when the possibility of non-Euclidean geometries was proven in the course of the nineteenth century, showing, among other things, that the shortest path may not be unique. Indeed, the 'shortest path' and 'the straightest path' can be defined independently. The elaboration of the mathematical formulation of these ideas provided the representational basis for the revolutionary reformulation of space and time in terms of a dynamic chrono-geometry and the inertio-gravitational field.

On the basis of this book, similar strands connecting the different forms of spatial knowledge could be sketched out. For this one would start from other elementary structures of spatial thinking, such as three-dimensionality, the distinction of the vertical direction, the definiteness and exclusivity of place, the continuity of motion, or the path-connectedness of space. In every case it is important to notice that the described cognitive structures not only change from one form of space to the other, they also have a different epistemic status in the context of each form of space. In particular, we have pointed out their different degrees of reflexivity. The recognition of this difference is not only important for an understanding of developmental dynamics, but also in the context of addressing long-standing questions of epistemology.

Take, for instance, the question of universal structures of cognition. We have argued that such structures are found in the forms of sensorimotor intelligence that are brought about by the natural conditions of spatial cognition. It is important not to conflate these structures with concepts, which would imply a claim of the universality of certain concepts. Concepts are cognitive structures on a higher level of reflexivity. While sensorimotor schemata may be described as unreflected, the genesis of concepts involves reflection on language, which is a cultural means for the external representation of knowledge. The differentiation according to degrees of reflexivity, together with the developmental approach, may in fact help to explain why certain concepts (such as *distance*, *void*, *place*, or *space* in the context of this book) display such striking similarities across cultures, while at the same time greatly varying in meaning depending on their cultural context. Their commonalities reflect their shared roots in universal, pre-conceptual structures, while their actual meaning is determined by their relation to the system of cultural artifacts, of which language is a part.

As explained above, we have here used the notion of *mental model* in order to describe similar cognitive structures functioning on different levels of reflexivity, which distinguish different forms of knowledge such as sensorimotor intelligence, practitioners' expert knowledge, or scientific theories. According to the approach presented here, scientific knowledge is therefore not detached from other forms of knowledge. It is part of a continuum of forms of knowledge acquired or produced in the context of different sorts of cultural activities. At the same time, it is distinguished from other forms of knowledge by (1) being a kind of systematically reflected, or theoretical, knowledge that (2) incorporates an ever increasing amount of experiential knowledge, to whose production it in turn contributes. Therefore, as should have become clear from the preceding chapters, the approach to a historical theory of knowledge presented here has consequences not only as regards questions of epistemology in general, but also more specifically as regards questions of the philosophy of science.

In particular it offers a perspective on scientific rationality and the history of science that does not put the two in direct opposition. Rationality is traditionally viewed as universal and ahistorical, and therefore claims of its historicity seem to threaten its validity altogether. History, on the other hand, is often viewed as contingent and irrational, and therefore claims of a logic of development, or even more so of progress, seem to imply an ahistorical element such as an underlying universal reason. It is from this sharp dichotomy that programs of a separation between rationality and history take their rationale, for instance when the rational reconstruction of the history of science is understood as the task to free a logical skeleton from the irrational flesh of the actual historical-psychological process of research and discovery, or when a historical but irrational context of discovery is distinguished from an ahistorical but rational context of justification. But of course there is no context of justification outside of history, strictly speaking, and the definition of knowledge as 'true, justified belief' itself turns out not to be ahistorical! All the failed attempts at applying an overly restrictive concept of rationality to real historical science have surely helped to nurture the idea that all rational justification of science has to be abandoned, which results in a complete relativism.

Here we have followed another way of conceiving of the rational reconstruction of the history of science: not as the task to free a logical skeleton from the irrational flesh of the actual historical-psychological process of research and discovery, but rather as the task to identify the rationality within the historical-psychological process. This approach implies a broader concept of rationality, a concept that cannot be reduced to formal logic. On the contrary, the reconstruction allows probing the horizon of possibilities of thinking given in the respective cultural and historical situation. According to this view, a separation of logical skeleton and historical-psychological flesh is not possible because rationality consists precisely in the application of the historically available patterns of thinking, i.e., structures of knowledge in action. Rather than seeking an ahistorical definition, knowledge may be understood as *experience* processed in such ways that it may serve to regulate and direct action and, on a higher level of reflection, to regulate and direct knowledge itself.

As was argued in this book, human spatial knowledge, be it the practical knowledge of expert navigators or the scientific knowledge of trained physicists, is saturated with experience. But the relation between experience and knowledge structures is historical-rational rather than universal-logical. While one may isolate certain parts of the architecture of knowledge and directly relate them to experiential knowledge, for instance when Newton's laws are related to the observation of projectile trajectories and planetary orbits, or when the affine structure of general relativistic spacetime is related to the experience of inertia, the relation is not one of logical necessity. There is no logically compelling induction from a raw experience of inertia, if it existed, to a particular mathematical representation. The relation is historical, but through history it is also rational: The processing of experience and the development of the representational means are the two sides of the historical process by which experience is rationally woven into the developing architecture of our knowledge.

In the context of all the forms of space discussed we have seen experiential knowledge being transformed into cognitive structures that serve as preconditions for further experience. There is thus always both an aspect of construction and an aspect of experience in the different forms of space. Both aspects are closely entangled, of course, because experience is always informed by cognitive structures already present in the mind, and, at the same time, it is experience that shapes the development of cognitive structures. One can thus say that there is no experience that is not structured by the mind, but there is also no mental structure that has not been shaped by experience. Our cognitive structures are the sediments of experience. But sedimentation is a historical process. This is why the understanding of the architecture of knowledge requires the historical analysis of its genesis.

Index

© The Author(s) 2016
M. Schemmel, *Historical Epistemology of Space*, SpringerBriefs
in History of Science and Technology, DOI 10.1007/978-3-319-25241-4

Printed in the United States
By Bookmasters

Printed in the United States
By Bookmasters